To Simon,

Happy Birtha

With mu

Angela

x x   06.03.06

OFFICER OF THE

MADRAS HORSE ARTILLERY

1817

From a Photograph of the Statuette the property of

THE OFFICERS' MESS

"J" BATTERY, ROYAL HORSE ARTILLERY

1898

THE
STORY OF
"J" BATTERY,

ROYAL HORSE ARTILLERY

(FORMERLY A TROOP, MADRAS HORSE ARTILLERY)

COMPILED FROM PRIVATE AND OFFICIAL RECORDS

BY
MAJOR GUILBERT E. WYNDHAM MALET,
CAPTAIN OF THE BATTERY 1875-79

The Naval & Military Press Ltd

published in association with

FIREPOWER
The Royal Artillery Museum
Woolwich

Published by
The Naval & Military Press Ltd
Unit 10 Ridgewood Industrial Park,
Uckfield, East Sussex,
TN22 5QE England
Tel: +44 (0) 1825 749494
Fax: +44 (0) 1825 765701
www.naval-military-press.com

in association with

FIREPOWER
The Royal Artillery Museum, Woolwich
www.firepower.org.uk

*In reprinting in facsimile from the original, any imperfections are inevitably reproduced
and the quality may fall short of modern type and cartographic standards.*

PREFACE TO THE FIRST EDITION.

THIS record of the services of J Battery, Royal Horse Artillery, was compiled by me at the request of the Officer Commanding the Brigade, as expressed in the following

Brigade Order by Colonel D. S. GREEN, C.B.,
Commanding D Brigade, R.H.A.

" DUBLIN, 14*th Sept.*, 1875.

" Officers Commanding Batteries are requested to collect all the information they can, from any available source, of the Records and Histories of their Batteries, and forward them to this Office on or before April next.

" The Commanding Officer hopes that Officers Commanding Batteries will do their utmost to have the Histories of their Batteries complete from the date of their formation up to the present time.

" The change of the designation of THE BATTERY, with the date, should be particularly noted, as also

" The names of the Officers who have commanded THE BATTERY from time to time ;

" The names of any Officers, Non-Commissioned Officers, or Men, who have distinguished themselves while serving in THE BATTERY ;

" The different Campaigns and Engagements, naming the General or other Officer under whose command THE BATTERY has served.

" Any General Order which reflects credit on THE BATTERY should also be copied and attached.

" The History should be supported by authority which should be quoted.—By order.

" (Signed) W. H. MONTGOMERIE.

"*Captain and Adjutant, D Brigade, R.H.A.*"

Many of the incidents recorded in the History have been, by permission, taken from the History of the Madras Artillery by the late Major-General P. J. BEGBIE, of that Regiment.

I take this opportunity of expressing my cordial thanks to General F. F. WHINYATES, Colonel-Commandant Royal Horse Artillery, Colonel J. WYNCH, Major-General J. D. MEIN, Colonel G. B. B. HOLMES, and other Officers who have served in THE BATTERY, also to Major-General P. ANSTRUTHER, C.B., R.A., for the assistance which I have received from their personal testimony. My thanks are also due to the Military Secretary at the War Office, to the Secretary of the Record Department at the India Office, and to the Secretary of the Royal Artillery Institution, Woolwich, through whose courtesy I have gathered much useful information.

BALLINCOLLIG, *July*, 1877.

PREFACE TO THE SECOND EDITION.

Twenty years have passed since the First Edition of THE BATTERY History was published. Many changes have taken place during that period. Fourteen Officers who served in THE BATTERY have gone to their rest, amongst them the sole survivor of the Battle of Mahidpur. The designation of THE BATTERY, too, has been changed for the eleventh, and, I trust, for the last time. At the request, therefore, of the present Commanding Officer, Major J. M. S. BRUNKER, I have much pleasure in bringing the History up to date, and have taken the opportunity to introduce new matter connected with THE BATTERY ; for, though no longer in the Service, the happiest recollections of my military career are associated with the five years I passed in it.

I am convinced that the most powerful incentives to a proper *esprit de corps* arise : (1) From a firm conviction that THE BATTERY to which we belong is the smartest Battery in the Regiment ; (2) From a proud and reverent appreciation of the glorious traditions of the past and the gallant deeds of those who have gone before ; (3) From a determination to emulate those deeds, and preserve those traditions in every possible way untarnished.

I may here mention that there are two Crimson Silk Trumpet Guidons in THE BATTERY, for use on special occasions. On each are embroidered the Crown within a Laurel Wreath, superscribed JAVA, NAGPORE, subscribed MAHIDPORE, in escrolls.

May this record help to keep those points before each and all of us, and tend to make THE BATTERY in the future, as it has ever been in the past, a credit to the British Army, our Country, and our Sovereign.

G. E. W. MALET,

Major, late R.H.A.

JUNIOR UNITED SERVICE CLUB,
PALL MALL, S.W., *March*, 1898.

PREFACE TO THE THIRD EDITION.

In conjunction with Captain C. A. Sykes I have completed The History of The Battery to the present date. The details of the South African Campaign have been taken from diaries written up immediately after the events as they occurred. The losses sustained by The Battery are accurately recorded, but those inflicted on the enemy could seldom be estimated, though at times they must have been considerable.

This part of The History is left in diary form, and the main facts only recorded, as it is practically a digest of services. There is no space to describe the individual experiences of the personnel, but it should be recorded that the men met the dangers and privations with extraordinary cheerfulness. This can be understood for the earlier portions of the campaign, when fighting was frequent; but to maintain the same spirit throughout the later phases, when subjected to constant sniping, continuous night marching, and consequent loss of sleep, reflects the greatest credit on all ranks.

P. H. ENTHOVEN,

July, 1903. *Major, R.H.A.*

CHANGES IN DESIGNATION.

1756   Madras Galloper Guns.

1805   The Troop of Horse Artillery.

1806   The Squadron of Horse Artillery.

1809   Rudyerd's Troop of Horse Artillery.

1818   Bonner's Troop of Horse Artillery.

1824   Conran's Troop of Horse Artillery.

1825   " A " Troop, Madras Horse Artillery.

1861   " A " Battery, 3rd Royal Horse Brigade.

1864   " A " Battery, " D " Brigade, R.H.A.

1877   " F " Battery, " B " Brigade, R.H.A.

1889   " J " Battery, R.H.A.

HISTORY OF "J" BATTERY

ROYAL HORSE ARTILLERY.

THE system of Horse Artillery, which has increased in so
marvellous a manner the mobility of an Army in the Field, is,
comparatively speaking, a modern development of warfare.
It has been stated on good authority that Horse Artillery
Galloper guns were first used by the English in India in
1756   1756.* From that time up to the close of the last century
they were gradually introduced into Europe,† viz., by Russia
in 1757, Prussia in 1759, Austria 1783, France in 1791, and
England in 1793.

The exact date on which the Madras Horse Artillery was
originally organized is unknown ; but even if the year 1756
cannot be claimed as the year of its birth, it is evident that
" J " Battery, Royal Horse Artillery, its lineal descendant,
may reasonably be considered to have commenced its existence
at some period between that year and 1767, for the Regimental
Records inform us that during the war with Hyder Ali of
1767   Mysore, 1767-69, it was usual to attach a Division of 2
Galloper guns of the Madras Flying Artillery to each
Regiment of Cavalry.

1780   On the 10th of September, 1780, a force of 3,800 men, with
10 Galloper guns, under the command of Colonel Baillie,
while on the march to join the main Army under General
Munro, was surprised and annihilated by the Mysore Forces
(80,000 men and 60 guns), under Hyder Ali and his son Tipu,

---

* Duncan's History of Royal Artillery.   † Rees' Cyclopædia.
B

at Polipur.   Amongst the killed was Lieut. Winter, of the Madras Horse Artillery.

1781   This disaster led to the first Mysore war*, and on the 16th of January, 1781, an Army was despatched from Madras under General Sir Eyre Coote, consisting of 800 Cavalry, 6,400 Infantry, and 60 guns, including a Troop of Horse Artillery, equipped with 8 3pr. guns.   On the 1st of July the Battle of Porto Novo was won, a victory upon which the whole fortune of the Madras Presidency turned, and at the second battle of Polipur, on the 27th of August, the enemy, under Hyder Ali and Tipu Sultan, were defeated with heavy loss, and the disastrous defeat of the previous year was avenged on the very scene of the disaster, and over the still unburied bodies of the slain.

Captain Innes Munro† also, in his history (p. 235), has recorded that two light field pieces drawn by horses were, during this war, attached to every Regiment of Dragoons and Native Cavalry.

1790   The Madras Galloper guns, under Lieuts. C. Donaldson and J. Neilson, were attached to the 19th Light Dragoons (Floyd's Regiment) in the second Mysore war, 1790-92.

The first action of mportance took place on the morning of the 13th of September. when a reconnoitring party of the British Cavalry met the vanguard of Tipu's Army at the Gazalhatti Pass, in which Colonel Floyd, with the 19th Light Dragoons and the Galloper guns of the Madras Horse Artillery, repulsed the enemy with the loss of 400 killed. This was followed by the Battle of Veládi on 16th September, the siege and defence of Bangalore, 7th to 19th March, 1791, the Battle of Arikera on the 13th of May, and other battles during the campaign under Lord Cornwallis, which ended on
1792   the 24th of February, 1792, when peace was declared before the walls of Seringapatam.

On the conclusion of the war, a Committee of Artillery Officers was assembled at Madras to consider and report upon the Equipment of Field Artillery.   As no mention appears to have been made by this Committee of the Galloper guns, it

---

* Stubb's History of the Bengal Artillery.
† Munro's History of the Mysore War.

was probably considered unnecessary for an Artillery Committee to consider the organization of a Corps which was at that time attached to the Cavalry.

1797   The first note of a change was sounded in 1797, in which year the Commandant of the Madras Artillery, Colonel, after wards Major-General, Thomas Geils, advocated the formation of the present system, but, much to the regret of the Officers of the Regiment, his excellent suggestions were not then carried out.

1799   In 1799 war was again declared with Mysore, and the Galloper guns of the Madras Horse Artillery, attached to the 19th Light Dragoons, and commanded by Lieutenant J. Limond, of the Madras Artillery, were constantly in action throughout the campaign, which terminated in the siege and capture of Seringapatam, under General Harris, on the 4th of May in that year.

Our defeat at the first Battle of Polipur was found thus recorded on a wall in Tipu's Palace :—" Killed 30 Officers and 155 European Soldiers ; wounded 41 Officers and 242 European Soldiers."

1801   In the Poligar war, 4 Galloper guns of the Madras Horse Artillery were attached to the 1st Madras Light Cavalry at the storming and capture of Punjalum-Coorchi, on the 23rd of May.

The storming party consisted of one company 74th Regt., the Grenadiers, and one company 77th and 5 companies of N.I. The enemy having been driven out, the Cavalry and Horse Artillery pursued and captured 600 prisoners.

The British loss in the assault was 43 killed and 172 wounded ; 450 of the enemy were killed.

The other battles during the campaign were Patiànur, 10th June, Ookur, 28th July, Sherewèle (the Murdu's capital), 30th July, and Peramáli, 17th September. Peramáli was a stone fort built on a spur of a rocky hill, the walls 40 feet high with a fortified Pettah outside.

The attack was made after a night march of fourteen miles over country flooded by heavy rain, and was entirely successful, the advance being covered by the Horse Artillery fire *

* " Welsh's Reminiscences," Vol. I., p. 75-119.

1803　　The Madras Galloper Guns were again sent on active service in the Mahratta Wars of 1803 and 1804, and are mentioned as having been present at the following engagements:—

> 1803.　First Mahratta War, Feb.-Dec.
> 　　8th Aug.　Ahmednagar.
> 　　21st Sept.　Assaye.
> 　　29th Nov.　Argaum.
> 　　15th Dec.　Gawilghur.
> 1804.　Second Mahratta War.
> 　　8th Oct.　Lasulghaum.
> 　　21st Oct.　Galnah.

1805　　On the 4th of April, 1805, a General Order was published giving effect to the suggestions made by Colonel Geils, and forming a nucleus of that independent system which, from that time to the present, has been considered by the highest military authorities to be the most suitable for the organization of Horse Artillery, although Galloper guns continued to be attached to Cavalry Regiments in Madras until the conclusion of the Mahratta war in 1818.

On the 8th of April, 1805, the following Officers were appointed to THE TROOP:—Captain-Lieutenant John Noble, Lieutenants W. M. Burton, G. W. Poignand (Adjutant), E. M. G. Showers, W. Shairpe, H. T. Rudyerd, R. B. Procter, and Assistant-Surgeon Ainslie.

The dress of the Officers, Non-Commissioned Officers and Men was as follows:—

> *Dress Jacket.*—Dark blue. Collar and cuffs, scarlet. Three rows of Horse Artillery buttons; laced from button to button—Officers with gold lace, Men with yellow braid; edged all round with the same. False pockets, embroidered at the sides.
>
> *Helmet.*—Grecian. Brass crest with scarlet mane. Headpiece covered with dark blue cloth, edged with scarlet. Ornaments, tiger's head and paws.
>
> *Breeches.*—White leather.

Boots.—Long. Steel spurs—strap and buckle.

Sword.—Broad. Steel scabbard, gold cord sword knot.

Sword Belt.—Buff leather for Men—gold lace for Officers.

Sash.—Officers—Crimson silk with gold cold and tassel.

Pouch.—Black leather.

Pouch Belt.—Buff leather. Gilt buckle and mountings.

Undress Jacket.—Same as dress jacket, but without the lace from button to button.

Overalls.—Blue cloth, slashed at waist with gold lace.

Boots.—Dress—Hessian, with gold tassel.

Caps.—Blue cloth—broad flat top—ornamented with gold cord.

For evening dress, Officers wore the dress jacket, leather or web breeches, and Hessian boots. They also had a short curved sword with brass scabbard, for undress.

THE TROOP, under its gallant Commander, speedily gained a reputation which justified the authorities in the step which **1806** they had taken, and induced them, on the 6th of January, 1806, to publish the following General Order, giving effect to a considerable augmentation:—

" The Horse Artillery, which was established by General Orders on the 4th of April, 1805, shall be denominated the Squadron of Horse Artillery, and shall be mustered, paid, and returned as such, from the 1st of January, 1806.

" The European non-commissioned rank and file shall be denominated the 1st half squadron.

"The Native Commissioned, non-commissioned rank and file shall be denominated the 2nd half squadron.

" To the 1st half squadron shall be posted 1 Captain-Lieutenant, 3 Lieutenants, 1 Assistant-Surgeon.

" To the 2nd half squadron, 3 Lieutenants."

The armament of THE SQUADRON consisted of two 12pr. guns; two 5½in. Howitzers, and ten 6pr. guns.

On the 27th of December, the following order was published after an Inspection of THE SQUADRON by the Commander-in-Chief :—

" HEAD QUARTERS, 27th Dec., 1806.

" The Commander-in-Chief derived extreme satisfaction from his inspection of THE SQUADRON of the Horse Artillery this morning. It was peculiarly gratifying to Sir John Craddock, who witnessed the birth of this force, to observe the rapid progress it has made towards perfection ; and His Excellency returns his warmest acknowledgments to Colonel Bell, Commandant of Artillery, for the pains he has bestowed upon this infant institution, and requests that he will communicate to the Commanding Officer, Captain Noble, and the other Officers of THE SQUADRON, how sensible he is that its creditable appearance in the Field, its regularity and good order in quarters, could not be effected without the most laudable and unanimous exertion.

" His Excellency entertains the warmest hopes that the Honourable Court of Directors may view the Establishment in so favourable a light that an extension will be directed, and that, in its due proportion, an arm which has had so powerful an influence upon the warfare of the present day in Europe, will not be denied to the Military character in India."

1808   On the 15th of March, 1808, Captain-Lieutenant C. Gahagan and Lieutenant E. J. A. Driffield were appointed to THE SQUADRON on augmentation.

On the 28th of December THE SQUADRON, in conjunction with 2 Regiments of Cavalry and 12 Companies of Infantry, was engaged at the Battle of Amulnair, in the Candeish, effecting the complete rout of a rebel force nearly 5,000 strong. The Governor-General expressed in General Orders his high sense of the zeal and judgment shown by Colonel Doveton, who commanded the force, and of the distinguished perseverance, exertion, and gallantry of the troops under his command, which was the more commendable, they having previously accomplished a forced march of nearly 100 miles in 48 hours.

1809   On the 27th of January, 1809, the denomination of THE SQUADRON was changed to that of "The Corps of Horse Artillery," which was formed into 3 Troops, 2 European and 1 Native, each being named after its Commanding Officer and equipped with eight 6pr. guns. (General Order dated 1st Feb.) Captain Noble was appointed Commandant of THE CORPS, Captain Rudyerd took command of THE TROOP, which was then quartered at Jaulnah, and Lieutenant Driffield became Adjutant. The establishment of each European Troop was 1 Captain-Lieutenant, 5 Lieutenants, of whom 1 was Adjutant, 1 Quarter-Master, 6 Sergeants, 4 Farriers, 6 Corporals, 2 Trumpeters, 12 Gunners,* 120 Matrosses.†

Lieutenant Walter Shairpe died at Mysore on the 25th of May in this year. Lieutenant R. B. Procter was appointed Quarter-Master of the Corps of Horse Artillery on the 3rd of July.

Captain-Lieutenant Poignand, Senior Officer of THE TROOP at Jaulnah, was one of seventeen Officers of the Madras Army to whom was given the option of leaving the Service or being tried by Court Martial (General Order, 25th Sept., 1809), " for joining a combination against the Government, in signing a paper addressed to the Right Honourable the Governor-General, purporting to be a remonstrance in the name of the Army against the acts of the Government, and for refusing to repudiate their conduct." He elected to leave the Service, and was accordingly struck off the Regiment on the 5th of April, 1810.

There were circumstances, however, in Lieutenant Poignand's case which led the Court of Directors to restore him subsequently to his rank in the Service, and he was therefore re-gazetted.

" Having taken into consideration the case of Captain Poignand, who preferred accepting the alternative of dismissal from the Service to being tried by Court Martial; also the

---

* The Gunners were especially selected for the care of the gun and ammunition, and appear to have fulfilled many of the duties which now fall to a Sergeant.

† Now Gunners and Drivers.

submission and penitent acknowledgments of that Officer, and the punishment he has suffered, WE do resolve to extend our lenity to him, and accordingly restore him to his rank in the Service."—(*Extract from General Orders dated 3rd June, 1814.*)

Troop-Serjeant-Major Brunton was appointed Quarter-Master of the Galloper guns of the 22nd Light Dragoons, under command of Captain Blundell of that Regiment, in August.

1811 On the 18th of April, 1811, THE TROOP under the command of Captain Rudyerd, the other Officers being Lieutenant Harrison, Lieutenant C. W. Black, Lieutenant and Adjutant E. J. A. Driffield, embarked at Madras for active service in Java, with the expedition under Sir S. Auchmuty, against the French under General Janssens. The British Army, 10,700 men in all, disembarked at Chillingching, at the mouth of the river Maraudi, about ten miles from Batavia, on Sunday, the 4th of August, and THE TROOP formed part of the advanced force which occupied Batavia on the 8th of August. On the 10th of August the Army advanced, and Colonel Gillespie's Division, to which THE TROOP was attached, attacked and defeated the French advanced Division under General Jumel at Welterwreeden, and captured the Arsenal there, which contained 300 guns. During the battle Lieutenant Driffield was mortally wounded, and died soon afterwards. Lieutenant Harrison succeeded him as Adjutant, and Lieutenant E. S. Munro was gazetted to THE TROOP. Colonel Gillespie, in his despatch dated 11th August, especially praises Captain Noble and the Officers and Men of the Horse Artillery for their gallant behaviour in the Battle of Welterwreeden.

The next operation in this war was the Siege of Cornelis, described in "Thorn's History of the War" as "an entrenched camp, defended by two rivers, with a number of redoubts and batteries, the circumference of the fortified lines comprising nearly 5 miles, defended by 280 pieces of canon and 17,300 men." Materials for the siege having been collected, the British Army broke ground on the night of the 20th of August. On the 22nd the enemy made a sortie which was repulsed. In the action Lieutenant Munro, of THE TROOP, was shot through

the arm, and it had to be amputated. On the 26th of August
the Commander-in-Chief determined to storm the works, and
at midnight THE TROOP, with Gillespie's Division, marched
off, making a long *detour* and attacking the enemy in flank at
daybreak. The feint was the signal for the general advance,
and after some hours' fighting, the French abandoned their
lines and retreated in the direction of Samarang with the loss
of 6,000 prisoners.

General Janssens fled to Buitenzorg, 20 miles distant, which
he attempted to hold, but was driven thence by the hot
pursuit of the Cavalry Brigade, in which THE TROOP, which
was attached to the 22nd Light Dragoons, joined.

He eventually escaped to Chéribon with the remainder of
his force, and thence to Salatiga, where he surrendered on the
18th of September.

1812   A detachment of THE TROOP sailed from Batavia on the
20th March, 1812, with the expedition under Colonel Gillespie,
to Palembang, a Dutch settlement on the Island of that name,
which had been treacherously attacked and destroyed by the
Sultan of Sumatra. It took part in the capture of Borang on
the 22nd and of Palembang on the 25th of April, when 249
guns were captured, and in the subsequent operations for the
dispersion of the rebels, and returned to Batavia on the 1st of
June. On the 6th of that month THE TROOP marched to
Samarang, and was engaged in the attack and capture of the
Crattan Fort, and the Battle of Djoecarta, on the 20th of June.
Lieutenant C. W. Black distinguished himself by blowing
open the gate of the Fort with one of his guus.

During these operations the country was so beset by the
enemy that it was impossible to keep open communication
with a force under the command of Colonel McLeod, of the
Royal Artillery, consisting of one Battalion of Infantry and
a Detachment of Captain Napier's* 10th Company, 6th
Battalion (now No. 3 Company, R.G.A.), under Lieutenant
Colebrook, R.A., which was *en route* from Salatiga to join
Gillespie. Even a native could not pass undiscovered nor
escape being murdered. In this dilemma, Gunner John

---

* Father of the first Lord Napier of Magdala.

O'Brien, of THE TROOP, volunteered to ride a day's march through the enemy's country with the dispatches for McLeod's detachment. Captain Rudyerd having granted his request, allowed him to select any horse he fancied, and recommended him *to make himself as light as possible, and to gallop when in trouble.* Gunner O'Brien accordingly started on his dangerous mission on the 18th of June, accomplished his object, and returned unhurt. The Commander-in-Chief, ever attentive to individual merit, rewarded this brave soldier with a gold medal, specially struck in commemoration of the exploit ; a reward, I imagine, for a private soldier, unique in military history. After the fall of Djoecarta the following General Order was published :—

" The conduct of Gunner John O'Brien, of the Horse Artillery, was particularly conspicuous, he having performed an important point of duty under circumstances of the greatest personal hazard, and he therefore merits public approbation."

The following paragraph also appears in the same Orders :—

" Captain Rudyerd, of the Horse Artillery, was conspicuous for gallantry and zeal. This valuable Corps is always conspicuous when its services are required."

1813   THE TROOP remained in Java until the following year, when it returned to Madras, arriving on the 13th of July, 1813. Soon after its arrival at Head Quarters the following Regimental Order was published by Colonel Robert Bell, Commandant of Artillery :—

" His Excellency Lieutenant-General the Hon. John Abercromby, Commander-in-Chief, having forwarded to the Commandant of Artillery a gold medal presented by Major-General Gillespie, Commanding the Forces in the Island of Java, to John O'Brien, Gunner of the Horse Artillery, as a mark of his approbation for brave conduct there on the 18th June, 1812, when he gallantly volunteered, at the hazard of his life, to carry an order through multitudes of the enemy, and on which depended the safety of a detachment of the Royal Artillery, Colonel Bell has most sincere gratification in being appointed to present this most honourable mark of approbation to John O'Brien, now Sergeant of the Horse Artillery, in the presence of the Corps of Artillery at this Establishment. Occasions can but seldom offer for such consummate gallantry, but Colonel Bell has a confident hope that all those who now witness the ceremony may be stimulated by the example of John O'Brien to a willing and zealous exertion of their best services whenever they may be required. His Excellency the Commander-in-Chief, in compliance with the request of Major-General Gillespie, permits John O'Brien, of the Horse Artillery, to wear the medal at all times."

" S. THOMAS' MOUNT, *17th October, 1813.*"

This Order was read, and the medal presented to Sergeant O'Brien, in the presence of the whole of the Artillery at St. Thomas' Mount, paraded for that purpose.

1815   The beginning of the end of the Cavalry Galloper guns was announced in the following Order :—

"The Gallopers of the Cavalry Regiments of the Subsidiary Force are brigaded ; and Lieutenant James of the 7th Regiment of Light Cavalry is appointed to the charge of them. *They are in all situations to be considered as under the orders of Major Noble Commanding the Corps of Horse Artillery*" (General Order, C.C., 1st April, 1815).

On the 4th of June Major Noble received the distinction of C.B.

From the 31st of Jan. to the 8th of June, 1815, THE TROOP was with the Army of Reserve. The Rocket Troop, which was raised in this year, was attached to THE TROOP under Captain Rudyerd's command.

1817   In April, 1817, the Mahratta war broke out, and THE TROOP was ordered from Madras to Secunderabad. On arrival at that station it was attached to General Sir Thomas Hislop's Division of the Army of the Deccan, and both Officers and Men were delighted at finding themselves again associated with their old friends of the Java campaign in the 22nd Light Dragoons. They left Secunderabad at the end of

August and crossed the river Poornah on the 6th of October. The following extract from a letter by a Staff Officer who was present when the Cavalry Brigade effected the passage of the river, shows the energy and pluck of THE TROOP :—

" The ford was reported practicable at noon on the 6th of October, and that confidence, known to men who are determined to succeed, effected the passage of a rapid stream, the Cavalry with wet seats, the limbers of the Horse Artillery under water, and without a single accident, although the guns came down the centre of the river upwards of 2 furlongs. It almost made the young Commandant believe that his Artillery could swim as well as fly ! "—in allusion to the designation, " The Flying Artillery," by which THE TROOP was popularly known in the Service.

Hislop's Division reached Nandair on the Godavery on the 16th of October, and joined Sir John Malcolm's Division of the Army of the Deccan on the 10th of November. The Army advanced at the end of November, and on the 21st of December the Battle of Mahidpur was fought, in which the Mahratta forces under Holkar were defeated with the loss of 3,000 men and 63 guns. A description of this great battle could not be comprised in this small work, nor, indeed, would it be necessary, its details having been so faithfully recorded by the historian of the war. THE TROOP suffered severely in the action. Covered by the Rocket Troop, it took up an advanced position after crossing the river Sipra under a heavy fire. In this position it remained, supporting the attack. One limber was blown up, many horses were killed, and 3 guns were totally disabled. Lieutenant J. J. Gamage and Quarter-Master Griffin were wounded. Sergeant Fanning, Gunners Flannagan, Flinn, and 2 Matrosses were killed, 10 Non-Commissioned Officers and Men were wounded. 3 Officers' chargers and 35 Troop horses were killed. The following extract from the General Orders by the Commander-in-Chief concerns THE TROOP :—

" The conduct of the Light Artillery Brigade, under command of Captain Rudyerd, was such only as the Commander-in-Chief expected, but His Excellency could not fail to be particularly struck by the admirable coolness and steadiness with which THE TROOP passed the river and took up its position within point blank range of the enemy's guns ; and to its well-directed fire and promptness in opening it too much praise cannot be given. His Excellency laments that this valuable Corps should unavoidably have sustained so severe a cannonade, by which at one time the whole of its guns were disabled " (General Order, C.C., Army Head Quarters, 22nd December, 1817).

The result of this victory was that Holkar's possessions in the Candeish were ceded to the British, and Sir Thomas Hislop's Division was directed to receive the surrender of the **1818** various Forts. The Division accordingly crossed the river Tapti on the 20th of February, 1818, and on the 27th attacked and stormed the Fort of Talnair, which had refused submission, in spite of the treaty. In this action THE TROOP again received the thanks of the Commander-in-Chief for its services, and Captain Rudyerd and Lieutenant J. G. Bonner were especially mentioned in dispatches (General Orders, C.C., 28th Feb., 1818). One Non-Commissioned Officer of THE TROOP was killed, and two Gunners were wounded in the action.

On the 3rd of March THE TROOP recrossed the Tapti, and proceeded to Jaulnah. Soon after its arrival at that station it accompanied a force under General Sir John Doveton, K.C.B., which was sent in pursuit of Baji Rao and Appa Sahib, whose army was eventually dispersed at Sioni on the 17th of April. In this pursuit THE TROOP was called upon to march nearly 100 miles in two days ; a feat similar to that which it accomplished under Doveton in 1808. It was a severe test of the powers of endurance of both men and horses, and its successful performance after a hard fought campaign deserves to be placed on record.

In the attack on the Fort of Belgaum, on the 10th of April, a Sergeant of THE TROOP, who was doing duty with the Gallopers of the 23rd Light Dragoons, was killed. The Peishwar surrendered on the 3rd of June, and on the conclusion of hostilities THE TROOP returned to Jaulnah. Second Lieutenant John T. Kelly died here on the 29th of June.

In the autumn of this year Captain Rudyerd was appointed Superintendent of the Gun-carriage manufactory at Seringapatam, and was succeeded by Captain J. G. Bonner. Towards the end of December Lieutenant F. F. Whinyates, who had served in THE TROOP during the Mahratta War, was detached in command of some Foot Artillery guns with the force which was engaged at the siege and capture of Jilpy-Amnair on the river Tapti, from the 2nd to the 11th of January, 1819.

In March THE TROOP was present at the siege of Assirghur, under General Sir John Doveton. This Fortress crowns the summit of a detached hill 750 feet high, about 1,000 yards long, and 600 yards broad, and is only accessible at two points. THE TROOP was engaged in the siege from the 17th of March to the 9th of April, its Officers and Men being in the Batteries during that period without relief night or day. On the latter date Assirghur surrendered. It is interesting to note that, although the casualties on our side amounted to fifteen Officers and 208 Non-Commissioned Officers and Men killed and wounded, the siege of Assirghur was not considered of sufficient importance to entitle the troops to a medal. The endurance and bravery shown by the British in this arduous undertaking will, however, be none the less remembered ; and this siege will deservedly be considered one of the most brilliant actions of the war.

THE TROOP returned to Jaulnah in May.

On the 8th April a General Order was published, directing the designation of "Bombardier" to be substituted for "Gunner," and "Gunner" for that of "Matross," in the Regiment of Artillery. The alteration did not affect the pay hitherto drawn by these ranks. By the same Order the rank of Lieutenant Fireworker was abolished.

1821 In 1821 the Rocket Troop, which for six years had been attached to THE TROOP, was formed into a distinct Troop under the title of the Reserve Troop of the Horse Brigade.

1822 Wellington boots and blue cloth overalls strapped with leather were served out in 1822, in place of the leather breeches and long boots, which were done away with.

1823   Lieutenant R. S. Seton, who had served in THE TROOP since 1816, was appointed Adjutant of the Golundauz Battalion in February.

1824   On the 11th of November, 1824, THE TROOP marched with the expedition under Colonel Deacon, against Kittoor, where Captain Black's, afterwards "E", Troop of Horse Artillery had suffered a disaster resulting in the death of that Officer (who had served as a subaltern in THE TROOP in Java), 2 Lieutenants, many of the men, and the loss of some of the guns. Captain Bonner having been appointed Assistant-Adjutant-General of Artillery on the 3rd of September, the command temporarily devolved on Lieutenant Whinyates. Under this Officer it proceeded by rapid marches to Sholapore, where Captain George Conran, its new Commanding Officer, joined. The force assembled at Kittoor towards the end of November. After some hard fighting, the Fort surrendered on the 4th of December, and Black's guns were recaptured.

1825   The six Troops of the Brigade of Horse Artillery, which had been hitherto known by the name of their Commanding Officers, were in this year lettered from A to F. Four of them still exist, viz., A, now J ; B, now M ; C, now P ; and D, now R. E and F Troops and the Rocket Troop, raised in 1815, have been reduced.

    THE TROOP returned to Jaulnah on the 8th of February, 1825. Lieutenant Whinyates (Commissary of Ordnance at Jaulnah) was, in this year, promoted to Captain and appointed to the command of the D Troop (now R Battery, R.H.A.) which was then raised, 12 men and 12 horses being detached from THE TROOP to form the nucleus of the new Troop.

1826   THE TROOP left Jaulnah on the 3rd of January, 1826, for Kamptee, on being relieved by B Troop (now M Battery, R.H.A.), and arrived there on the 31st of January.

1827   On the death of Captain N. Hunter (who was in command of B Troop), on the 7th of May, 1827, Captain Conran was transferred to that Troop in order that he might hold the command of the Artillery with the Light Field Division of the Hyderabad Subsidiary Force, to which the B Troop was attached. He was succeeded in command of THE TROOP by Captain R. S. Seton.

1831 On the 26th of April, 1831, THE TROOP left Kamptee for Madras. Captain Seton having been transferred to a Native Troop, Lieutenant Best was in command on this march, and Captain John Wynch, its new Commanding Officer, joined THE TROOP on its arrival at Madras on the 31st of July.

1832 Lieutenant-Colonel Bonner, Inspector-General of Military Stores, who commanded THE TROOP 1818—24, was selected by the Indian Government to take charge of certain banners and war trophies won by British Troops in India, and deliver them to Lieutenant-Colonel Wilson to be placed in the Royal Military Hospital, Chelsea, in accordance with His Majesty's desire, expressed on the 20th of April, 1832.

1836 On the 11th of January, 1836, THE TROOP marched to Bangalore. Lieutenant-Colonel Noble, C.B., who had been its first Commanding Officer under the Regimental system, and who had been for 27 years Commandant of the Horse Artillery, died on the 16th of July, on board the Ship " Roxburgh Castle," on his way to India after furlough.

1837 On the 4th of April, 1837, Sergeant S. Wilson, of THE TROOP, was appointed Quarter-Master of the 52nd Regiment, N.I. From the 12th of April until the 27th of May THE TROOP was encamped with a Regiment of Cavalry at the Mysore Residency, where a disturbance was apprehended and prevented.

1838 In a report made by Sir H. Gough, K.C.B., dated the 9th January, 1838, the following paragraph appears :—

" The very high state of efficiency in which I found the A Troop of the Horse Artillery denotes the zeal and attention of Captain Wynch, and of all under his command,* and calls for my marked approbation."

This Officer was shortly afterwards appointed Assistant-Adjutant-General of Artillery, and Captain John Timins Baldwin received command of THE TROOP.

On the 9th July in this year the Right Honourable the Governor-General was pleased, under permission from the

---

* Lieutenants E. Brice, C. J. Cooke (Adjt.), J. K. B. Timins, and W M. Gabbett.

Honourable the Court of Directors, to confer on Deputy-Assistant-Commissary John O'Brien, who had distinguished himself when a Gunner in THE TROOP during the Java campaign, the rank of Lieutenant in the Veteran Battalion.*

On the 28th September THE TROOP left Bangalore for Sholapore, and arrived there on the 10th of November.

1839   Brigadier-General England, on the 30th July, 1839, reports, " In all points Captain Baldwin's Troop may be considered to be in a high state of order and efficiency."

1840   On the 13th August Major-General Sir James Limond, C.B., the last surviving Officer of THE TROOP as The Madras Galloper Guns, died at Boulogne-sur-Mer, France.

1841   On the 17th September, 1841, three guns of THE TROOP were detached on service in the Nizam's territory against some insurgent Arab and Rohilla tribes. They returned to Head Quarters on the 29th of September. Lieutenant J. K. B. Timins died at Kurnool on the 21st December.

1842   After an inspection of the Horse Brigade in November, 1842, the Inspecting Officer reports, " The A Troop, under Captain Baldwin, is as perfect in every respect as any Troop could possibly be."

1843   In November, 1843, the efficiency of THE TROOP was again highly commended in Orders.

1844   Towards the end of 1844 disturbances broke out in the South Mahratta country, and on the 5th of October three guns of THE TROOP, under Captain J. E. Maudesley, with Lieutenant B. W. Black, were sent, in conjunction with a force of 1,800 men under General Delamotte, C.B., to the disturbed districts. The force marched from Samunghur on the 21st of October. On the 7th of November, while on the march between Wittoor and Badraghur, an alarming accident happened to one of the wagons, the horses of which, being frightened at some obstruction in the road, placed by the

---

* Lieutenant O'Brien died at Vizagapatam on the 18th of November, 1842. His widow survived him till 1864. Shortly before her death a box containing her husband's gold medal (See p. 10) was stolen and never recovered. The melting-pot was doubtless the ignominious end of this priceless record of a gallant deed.

C

enemy, swerved and backed it over a steep bank. Wagon, limber, and horses went over. The limber was broken to pieces, but, strange to say, neither men nor horses received any serious injury.

On the 10th of November the force captured Badraghur, and on the 20th marched to Purnalla, arriving on the 25th. This fort was stormed and captured on the 30th of November, the advance of the storming party being covered by the guns of THE TROOP. The neighbouring Fort of Powenghur was taken next day. On the 28th of December General Delamotte published the following Division Order :—

" The Detachment of A Troop, of the Horse Artillery, being no longer required with the Field Force, the Major-General Commanding takes this opportunity of acknowledging its valuable services, which have been of no ordinary nature during the time it has been under his immediate command. Its uniform, steady, soldier-like conduct, reflects great credit on Captain Maudesley and Lieutenant Black."

1845 The half-troop returned to Sholapore on the 16th of May, 1845.

1846 On the 11th of April, 1846, Captain Baldwin died at Sholapore, and Captain Charles Lancaster succeeded him in command of THE TROOP.

1847 From the 10th of March to the 10th of May, 1847, THE TROOP was on the march from Sholapore to Madras. It remained at St. Thomas' Mount until the 4th of September, 1851, on which date it proceeded to Jaulnah, and arrived there on the 25th of November.

1848 On the 18th of March Lieutenant A. T. Cadell, of THE TROOP, was appointed A.D.C. to His Excellency the Governor of Madras.

1849 Lieutenant G. S. Elliott died at Jaulnah on the 21st August.

1852 In July, 1852, Captain Lancaster was invalided, and was succeeded by Captain G. S. Cotter. Lieutenants A. V. Falls, W. C. F. Gosling, and R. Morton were Subalterns in THE TROOP, which was then quartered at S. Thomas' Mount.

1854   On the 2nd of March, 1854, Captain Cotter resigned, and was succeeded by Captain J. D. Mein, under whose command THE TROOP returned to Jaulnah.

1856   On the 1st of December, 1856, THE TROOP marched for Madras. It was detained at Secunderabad for about 6 weeks, pending the result of an application to Head Quarters, that it might be detained in that command on account of the threatening aspect of affairs in the neighbourhood of the district ; the request, however, was not granted, and THE TROOP proceeded on its march, and reached S. Thomas' Mount on the

1857   2nd of March, 1857.

The Indian Mutiny broke out in May, and on the 27th of June THE TROOP was ordered to proceed to Government House for the purpose of defending it against an anticipated rising of the native population. It remained there for five months, the men being accommodated in the Banqueting Hall. No important event occurred during that time, excepting the disarming of a regiment of Native Cavalry near Madras, on its march from Bangalore, when the Battery was present. The order was carried out sullenly, but without a murmur, so great was the moral effect of the 6 guns ready for action. On 19th November THE TROOP was ordered to Secunderabad, and arrived there on the 27th of December. On the march 80 horse-keepers and some of the native artificers deserted. This threw extra work on the men of THE TROOP.

1858   In January, 1858, THE TROOP left Secunderabad to join the Saugor Field Force in Central India on being relieved by the B Troop * from Trichinopoly. Soon after leaving Kampti it was joined by Captain Palmer's Light Field Battery and a Detachment of the 3rd Madras Light Infantry. It joined General Whitlock's force at Dummoo on the 20th of March. The Subalterns at this time were Lieutenants R. G. F. Henegan, J. Stewart, and D. D. Anderson. THE TROOP took part in the following engagements during the month of April, viz.: Jheegum on the 16th, Kubrai on the 17th, Banda on the 19th, and Jhounpur on the 26th. During the remainder of the campaign THE TROOP, with 2 squadrons of the 12th

* Now M Battery, R.H.A.

Lancers, under the command of Major Oakes, and a Regiment of Irregular Cavalry, was frequently engaged in hunting up rebel localities in the Rewah, Kimur, Chirkiri, Chuttirpore, Hamirpore, Bhopal, and Banda districts. This, and the fact that Banda is 1,100 miles from Madras, will give some idea of the distances travelled over by THE TROOP during the war. When at Mahoba on one of these expeditions it was suddenly ordered to Kirwi, a distance of eighty-four miles. Within thirty-four hours THE TROOP, with the Cavalry Brigade, arrived at Kirwi, and took part in the relief of that town on the 25th of December. The last important action that THE TROOP was engaged in was the storming of the heights of Punwari on the 29th of December, on which occasion it was taken up the Duddri Pass in support of the right attack. A Farrier of THE TROOP, who was in the action, writes to the Compiler, "The enemy thought that we were mad to try and get the guns up such a place." At Punwari Radha Govindhoo's Army was completely routed, and the campaign was over excepting the action of Keuti on the 4th of March, 1859, in which two guns of THE TROOP took part.

1859   After more than twelve months of incessant hard work, and seventeen under canvas, THE TROOP marched into Saugor on the 26th of April, 1859, to be quartered there. Its casualties had not been many, but several men died from sunstroke and exposure, amongst whom were Assistant-Surgeon Denman, Quarter-Master Harris, and Quarter-Master-Sergeant Horne.

1860   On the 11th of February, 1860, THE TROOP left Saugor for Kampti, and arrived there on the 10th of March.

1861   Captain Mein was promoted on the 11th of August, 1861, and was succeeded in the command by Captain A. T. Cadell, who, however, only remained with THE TROOP for three months. His successor was Captain G. B. B. Holmes.

On the amalgamation of the Royal, the Bengal, the Madras, and the Bombay Regiments of Artillery, which was carried out in this year, the designation of THE TROOP was changed, and it became A Battery, 3rd Royal Horse Brigade, on the 14th of October.

The Busby was substituted for the Grecian Helmet, which had been the head dress since 1805, and the uniform was assimilated to that of the R.H.A. In the equipment, shafts were substituted for poles, which had been the method of draught for over 100 years.

1863 On the 15th of December, 1863, THE BATTERY marched **1864** for Bangalore, and arrived there on the 24th of February, 1864. In this year its designation was changed to A Battery, D Brigade, R.H.A.

1866 In March, 1866, Captain Holmes was invalided, and was succeeded by Captain N. G. Campbell.

1868 On the 12th of July, 1868, Major J. Denton, of the Madras Invalid Battalion, died at Cannanore, where he held an appointment in the Ordnance Department. He served in THE BATTERY as Pole Driver, under Captain Rudyerd, throughout the Java campaign, and at the time of his death was probably the last survivor of that war. He was Quarter-Master of E Troop at the time of the disastrous attack on Kittoor on the 23rd of October, 1824, and was given a Commission for his distinguished gallantry on that occasion, when Captain Black and Lieutenant Dighton having been killed, and Lieutenant Sewell mortally wounded, he took the command, and, overpowered by superior numbers, fought his way with the remainder of the Troop out of the action, in which he himself was three times wounded.

On the 12th of December THE BATTERY arrived at Kampti from Bangalore.

1869 Captain Campbell was promoted on the 1st of September, 1869. His successor was Captain E. S. Milman.

1871 On the 15th of December, 1871, THE BATTERY left Kampti and marched to Secunderabad, arriving there on the 24th of January, 1872.

1872 On the 5th of July, 1872, Captains of Batteries were made Majors and 2nd Captains became Captains.

On the 1st of August Lieutenant-Colonel S. H. E. Chamier was appointed to the command, vice Milman, but THE BATTERY being under orders for England, Lieutenant-Colonel Chamier, who was at home, did not take over the command until its arrival. THE BATTERY marched from Secunderabad on the 31st of October, and embarked at Bombay on board H.M.S. *Serapis* on the 22nd of November. It disembarked at Woolwich on the 23rd of December.   Captain Rawlins, Lieutenants Gambier and Hutchinson, were the Officers with THE BATTERY on the passage home.

1873   In April, blue cloth breeches with scarlet stripe, long boots, and strapped spurs, were issued to THE BATTERY in lieu of booted overalls, which had been the equipment since 1822.

In June, Driver J. Bird, who was Wheel Driver of No. 4 gun, had a narrow escape.   His horse stumbled and fell when THE BATTERY was galloping past at an inspection, and his right ear was cut off by the near wheel of the gun.

1874   On the 10th of April, 1874, THE BATTERY left Woolwich for Ireland, arriving at Dublin on the 25th, *viâ* Birkenhead.

Driver J. Belchamber was awarded the medal of the Royal Humane Society for gallant conduct in jumping into a canal on the 10th of May, and rescuing from death a woman who had attempted suicide.

Lieutenant-Colonel Chamier, a very clever musician, took his degree as Bachelor of Music in the University of Dublin in the autumn, the degree being conferred upon him dressed, by special request, in full uniform.

1875   A handsome Eagle Lectern in the Garrison Chapel at Portobello Barracks, subscribed for by the Officers, N.C. Officers and Men of THE BATTERY, commemorates a sad accident in the Phœnix Park, at the Review on the 29th of May, 1875, by which Driver J. Meuris of THE BATTERY lost his life.   On the 20th of July THE BATTERY marched from Dublin for Ballincollig, arriving at that station on the 28th of July.

The trails were fitted with the experimental swivel pintle, designed to prevent the limber from upsetting over rough ground ; but as the guns constantly came into action upside down, the invention was not favourably reported on.

1876   THE BATTERY left Ballincollig on the 31st of May, 1876, under orders for Dublin for practice and, on its conclusion, marched to the Curragh for the drill season.

Lieutenant-Colonel Chamier left THE BATTERY, on promotion, on the 1st of July. THE BATTERY, under the command of Captain Malet, remained at the Curragh until the 16th of August, when the various Corps assembled there returned to winter quarters. On the march that day the horse ridden by Gunner C. Covell (one of the billeting party) bolted with him between Athy and Castle Comer, and, on passing some carts, knocked down and fell on a wagoner, who was killed. Covell escaped with a few bruises, and on arrival at Castle Cromer was arrested by the Civil Power. He was tried for manslaughter at Carlow quarter sessions in October, and was acquitted. Major A. M. Rawlins joined THE BATTERY on the 28th of October.

On the 1st of September Lieutenant Josselyn had probably one of the narrowest escapes from death ever known. He was standing on the platform at the Ballincollig railway station, and fainted as the train came in, falling between the platform and the engine. When the train stopped he was lying partly across the rail in front of the carriage next to the engine, severely bruised, but with no bones broken.

1877   THE BATTERY marched to Kinsale, for the annual practice, on the 12th of June, returning to Ballincollig on the 23rd of the same month.

On the 1st of July, 1877, the designation of THE BATTERY was changed to F Battery, B Brigade, R.H.A.

This was considered a suitable time by the Officers of THE BATTERY for inviting all Officers who had served in it to subscribe for the purchase of a piece of plate for the Mess, which should be a memorial of its brilliant services in Java and India.

The scheme was most favourably received, and the beautiful silver statuette representing an Officer of the Madras Horse Artillery, in the uniform of 1817, the year in which the Battle of Mahidpur was fought, forms the subject of the frontispiece of this History.

The Officer is supposed to have galloped to the front, and having found a suitable position for the guns, is giving the signal for THE BATTERY to advance.

On the basement, in front, is the following inscription :—

TO COMMEMORATE THE SERVICES

OF

"A" TROOP, MADRAS HORSE ARTILLERY,

FORMED IN 1805

FROM

THE GALLOPER GUNS,

ORGANIZED IN 1756

BY

THE HONOURABLE EAST INDIA COMPANY

IN THE MADRAS PRESIDENCY,

BECAME

"A" BATTERY, 3RD ROYAL HORSE BRIGADE, 1861,

"A" BATTERY, "D" BRIGADE, ROYAL HORSE ARTILLERY, 1864,

"F" BATTERY, "B" BRIGADE, ROYAL HORSE ARTILLERY, 1877.

PRESENTED

BY

OFFICERS SERVING AND WHO HAVE SERVED

IN

THE BATTERY.

1877.

In rear are the Arms and Motto of the Royal Regiment of Artillery. On the right side is a list of the actions in which THE BATTERY has been engaged since the introduction of the Regimental system, and on the left the names of the Subscribers.

1878　　On the 4th of March, 1878, THE BATTERY embarked at Queenstown on board H.M.S. *Assistance*, being relieved by B Battery, A Brigade, now " B," R.H.A., from Exeter. It disembarked at Plymouth on the 6th, after a very rough voyage, and arrived at Exeter on the 8th of that month.

In April, the Officers, Non-Commissioned Officers and Men of THE BATTERY raised a Subscription for a stained glass window to be placed over the altar in the Garrison Chapel at Ballincollig, the improvement of which was at this time sanctioned by the War Office, to be carried out by Captain Malet, in memory of his wife, who died on the 14th of September, 1877.

Her Majesty Queen Victoria having expressed a wish to see the silver Statuette belonging to the Officers' Mess, it was sent to Windsor Castle on the 27th of April for Her Majesty's inspection.

1879 Captain Malet was gazetted an Esquire of the Order of S. John of Jerusalem on the 23rd of June.

THE BATTERY carried out the annual practice at Oke-hampton.

1880 THE BATTERY subscribed for a window representing " The Annunciation," in Ballincollig Chapel, in memory of Mrs. Rawlins, wife of their Commanding Officer, who died in September, 1879.

1881 THE BATTERY arrived at Aldershot on the 5th of May from Exeter, and in June won the 2nd Prize at the Royal Military Tournament, at Islington, for precision in walking, trotting, and galloping with the 13pr. gun.

1882 THE BATTERY was placed in the 2nd Army Corps, and the armament reduced from 6 to 4 guns.

1883 Major Rawlins was promoted on the 14th of January, and Major and Brevet Lieutenant-Colonel W. Ward took over the command. On the 22nd of May THE BATTERY marched from Aldershot, *en route* to Okehampton, for the annual practice, returning to Aldershot on the 2nd of July.

1884 THE BATTERY marched to Canterbury in April, and the Annual Practice was carried out at Lydd in August.

Major and Brevet Colonel F. S. Le Grice was appointed to the command, vice Ward, promoted.

1885 The Annual Practice was held at Lydd in July. On the 29th of December THE BATTERY, under the command of Major R. Corbett, embarked at Portsmouth on H.M.S. *Jumna* for service in Egypt. It disembarked at Suez on

1886 the 15th of January, and marched to Cairo, relieving G Battery, B Brigade (now M), R.H.A., at Abbassiyeh Barracks.

1887 General F. F. Whinyates, Colonel Commandant, R.H.A., who served in THE BATTERY as a Subaltern, 1817-1824, and was present with it during the Mahratta War, died on the 22nd of January. With his record of 74 years' service, he was the Senior Officer of the Regiment at the time of his decease.

On the reduction of the Army of Occupation in Egypt, THE BATTERY was ordered home, and embarked at Alexandria, on H.M.S. *Assistance* (whose prodigious rolling powers had given anything but lively satisfaction ten years before), on the 1st of March, and after experiencing some rough weather, which compelled the ship to anchor for a few days in Corcubin Bay, on the north-west coast of Spain, disembarked at Woolwich on the 21st of March.

This was the fiftieth year of the reign of Queen Victoria. The Gunners of THE BATTERY were detailed to keep the ground at the Royal Visit to the East of London on the 14th of May, and again at the State Procession to Westminster for the National Thanksgiving Service, on the 24th of June. THE BATTERY marched to Aldershot, for the Royal Review, on the 9th of July, with their new 12pr. 7 cwt. B.L. guns, returning to Woolwich on the 12th of July. The Annual Practice was carried out at Shoeburyness.

1888 On the 23rd of June Major Malet, who was Captain of THE BATTERY 1875-79, was gazetted as a Knight of Grace of the Order of S. John of Jerusalem, and on the 8th of November was appointed by H.R.H. the Duke of Cambridge, Commander-in-Chief, to be Director of the Royal Military Exhibition, 1890, at Chelsea, which was followed by the Royal Naval Exhibition in the succeeding year. On the 4th of July the Military Exhibition was visited by Her Majesty the Queen. Both Exhibitions were in every way successful, and resulted in a surplus of more than £57,000 for Military and Naval charities. The total number of admissions was 3,275,444.

1889 The name of THE BATTERY was again changed on the abolition of Artillery Brigades, and it became " J " Battery in accordance with Regimental Order No. 89, dated 1st July. It marched to Aldershot for the Royal Review on the 1st of August, and, after a week in camp, was stationed at St. John's Wood Barracks, London. During the remainder of the summer THE

BATTERY on several occasions took part in Displays at the Royal Military Exhibition at Chelsea.

1891   On the 7th of April THE BATTERY was ordered to Woolwich, being replaced at St. John's Wood Barracks by the Royal Horse Guards. From the 14th of May to the 26th of July THE BATTERY was in camp at Aldershot. On the 12th of August Major W. L. Davidson succeeded Major Corbett in the command on the promotion of the latter to Lieut-Colonel.

1892   At the Annual Practice at Shoebury in September THE BATTERY won 1st prize in the Competition, scoring 300 out of possible 370 points.

1893   The Commanding Officer, with 1 Subaltern and 69 Non-Commissioned Officers and Gunners, marched to Aldershot in May, and formed the Ammunition Column during the manœuvres.

THE BATTERY was again successful at Shoebury, winning the 1st prize with 338 points out of 370.

1894   THE BATTERY left Woolwich for Aldershot on the 10th of May, and was encamped on Farnborough Common from the 14th of June to the 16th of July, when it went to Okehampton for the Annual Practice, returning to Aldershot on the 20th of August.

1895   The old system of pole draught (abolished in favour of shafts in 1861) was reverted to.

1896   Major E. Vaughan-Hughes was appointed to the Command on the 16th of February, vice Davidson, promoted.

1897   THE BATTERY was ordered from Aldershot to Woolwich, and arrived there, after the Annual Practice at Okehampton, on the 4th of June. On the 19th it took part in the " People's Procession " through the East End of London. Leaving Woolwich at 10 a.m., it marched through the Blackwall Tunnel to Victoria Park, and thence to the Mansion House, returning viâ the Tower Bridge and Deptford. It also formed part of the escort in the Queen's Procession through London on the 22nd of June, on the completion of Her Majesty's sixtieth year on the Throne of England. Leaving Woolwich at 6.30 a.m., THE BATTERY halted in Eaton Square at 8. Here a resident in the Square (Mr. L. Hall) entertained the Officers at breakfast. At 9.30 it took up its position in Pall

Mall until Her Majesty left Buckingham Palace, when the escort moved off on its route to S. Paul's. On the return from the City, through Southwark and Westminster, the six Batteries of Horse Artillery and six Regiments of Cavalry watered and fed in Hyde Park. All were most hospitably entertained by the Household Brigade, who provided refreshments for Officers, Non-Commissioned Officers, and Men.

THE BATTERY returned to Woolwich at 5 p.m.

On the 1st of July Major Vaughan-Hughes exchanged with Major J. M. S. Brunker of Q Battery.

A farewell dinner was given by past Officers to those now serving, at Hôtel Dieudonné, on the 20th of September. Present :—

Guests.

Major J. M. S. Brunker, Captain A. C. Connal, Lieutenants C. E. D. Budworth, J. A. Hobson, and G. N. Cartwright.

Hosts.

Lieut.-General S. H. E. Chamier, C.B., in the Chair. Major-General A. M. Rawlins; Colonels R. Corbett, W. S. Hebbert, J. E. Josselyn ; Lieut.-Colonels W. L. Davidson, W. L. Hutchinson, C. W. Thomson, L. Tillotson ; Majors W. E. Fairholme, W. F. L. Lindsay, C. Maberley, G. E. W. Malet ; Captains J. D. Barry, H. Calley, F. A. G. Y. Elton, F. W. Heath ; Lieutenant P. Sanderson, C.M.G.

On the 13th of October THE BATTERY embarked at Southampton on the s.s. *Dilwara* for India. Strength :— 5 Officers, 86 Non-Commissioned Officers and Men. Their destination being Bangalore, relieving " M " Battery, ordered home.

The *Dilwara* left Southampton on the 14th, and arrived at Bombay on the 4th November. On the 10th of the month THE BATTERY reached Bangalore, in which place it was first quartered in 1836.

1898 Annual practice was carried out at Nundidroog from the 1st to the 14th of January, the camp being one day's march from cantonments.

The second BATTERY dinner was held in London on 13th June.

1899   Captain A. C. Connal died at Cairo on the 2nd February, having been taken seriously ill on his way home.

The annual dinner took place on 17th June.

Major E. Vaughan-Hughes, a former Commanding Officer, died on the 30th August.

On Lieut. the Hon. G. F. Stanley exchanging in September, Major Brunker remained the only Officer who had left England with THE BATTERY.

On the 5th October 1 Shoeing Smith, 3 gunners, 3 drivers, and 13 horses left for South Africa with the Reserve Horse Contingent, and 3 months later THE BATTERY itself was on its way to the seat of war.

THE SOUTH AFRICAN WAR.

1899
Dec. 31
THE BATTERY was at Bangalore preparing for practice camp when the various reverses occurred in South Africa in December, 1899. On the 31st orders were received to mobilize and embark for the front. Major Brunker was promoted, and parted with his Battery at the station on **1900** January 8th, 1900, when it entrained for Bombay, receiving a **Jan. 8** splendid send-off.

Six 15pr. guns from the 40th Battery R.F.A. were taken, as the 12pr. 7 cwt. belonging to " J " Battery would have complicated the ammunition at the front.

Jan. 11 THE BATTERY reached the Western port on the 11th, when Major P. H. Enthoven assumed command. Fifty-four men and seventy horses from " L " Battery, Secunderabad, also joined, and the whole embarked on the s.s. *Ujina*, said to be the finest transport that had left the docks. She sailed the same afternoon.

The Officers with THE BATTERY were :

> Major P. H. Enthoven,
> Captain H. Corbyn,
> Lieutenant W. P. Monkhouse,
> Lieutenant V. R. Hine-Haycock,
> Lieutenant L. H. Delves-Broughton, and
> Surgeon-Major R. I. Power ;

with Sergeant-Major Cocks and Quarter-Master-Sergeant Gardner, Sergeant Farrier Dickson and Sergeant Wheeler Collinson, Sergeants Watson ("A" Sub-Div.), Giles ("B" Sub-Div.), S. Major (" C " Sub-Div.), Salmon (" D " Sub-Div.), Calvert (" E " Sub-Div.), Henson (" F " Sub-Div.), and the following Corporals, who were afterwards Sergeants : Eden, H. G. Kerry, and G. L. Kerry. Corporal Haynes was later in

1900 charge of the Kaffir mule drivers, and Corporal Gray was the senior Collar Maker. The total strength was 184, including Officers, with 209 horses, including 11 chargers.

After a quick and uneventful voyage Durban was reached **Jan. 25** on the 25th. Orders were received for THE BATTERY to proceed to Port Elizabeth, "A" Battery, which had just arrived, being retained. The disembarkation was carried out **Jan. 29** at the latter port on the 29th by means of lighters, but no casualties occurred. Camp was pitched just outside the town, **Jan. 31** and on the 31st the men received the Queen's chocolate boxes, and entrained for Orange River Station. Large crowds greeted THE BATTERY enthusiastically both at the station and as the train passed through the town. Cradock was reached **Feb. 1** at 4 p.m. the next day, and Naaupoort at 4 a.m. on February **Feb. 2** 2nd. Here (owing to the calibre of the guns) fresh orders were received, directing THE BATTERY to go to Rensburg.

Reaching this station, THE BATTERY detrained and marched at once to Slingersfontein, the extreme right of the line and some 13 miles from Rensburg, the headquarters of General French. An incompetent guide misled THE BATTERY, and added to the difficulties of the march.

The force on this flank, under the command of Major-General Clements, consisted of:

1 Squadron Inniskilling Dragoons,
1 Squadron West Australian M.R.,
1 Section 37th (Howitzer) Battery,
1st Battalion Royal Irish Regiment,
2nd Battalion Worcester Regiment,

nearly all the mounted troops having been secretly withdrawn to Rensburg, in course of conveyance to Orange River.

Feb. 4 On Sunday, Feb. 4th, a portion of THE BATTERY was in action for the first time. The Officer Commanding took the Right Section (Lieutenant Monkhouse) with an escort, and moving out before dawn to a hill in the vicinity of "Worcester Kopjes"— the outpost line held by that Regiment—shelled Kleintoran Hill occupied by the Boers. A burgher subsequently met on his farm in the Free State vouched for the accuracy of the fire.

1900
Feb. 5   On the 5th the Left Section under Lieutenant Hine-Haycock returned to Rensburg, from whence it was in action on the 9th and 10th.

Feb. 6   On the 6th Major Enthoven conducted a reconnaissance of Potfontein farm. Taking the Right Section, fifty Inniskillings and sixty Australian M.R. (who were in action for the first time) the "nine Boers" whom he was informed would be there proved to be some six hundred. A sharp engagement took place, one gun wheel showing twenty-eight bullet marks. Sergeant Watson was dangerously wounded, dying the next day, several horses were hit, and the cavalry suffered six casualties. The Boer ambulances were busy on the neutral

Feb. 7   ground during the night. The next day a similar reconnaissance showed that the enemy had evacuated the farm.

Meanwhile, the Left Section, under Lieutenant Delves-Broughton, had at daybreak on the 6th surprised the Boers near Worcester Kopjes while constructing a gun epaulment, and, according to two deserters brought in on the 8th, hit thirty-three out of the fifty comprising the working party. The range was but 2,400 yards.

Feb. 9   On the 9th it was found that the enemy had worked round the right flank and a patrol of twenty West Australians under Captain Moore, R.A. (attached to that corps) was almost surrounded on a kopje which it had occupied. THE BATTERY effectually protected the flanks for the whole day, and at dusk the riflemen were enabled to withdraw with a loss of one killed and one wounded. The four guns had fired 511 rounds, but had arrested the further enveloping movements of the enemy.

Feb. 10   On the 10th the camp was shelled by a long range gun on a hill to the north and an artillery duel took place between it and a section in strong emplacements in the plain, some 3,000 yards from camp and 4,000 from the enemy. The Boers during the day fired shrapnel filled with French nails, but did no damage. On the other hand, their officer, apparently a German, was put out of action.

Feb. 11   Sunday, the 11th, no response was made to the few rounds fired at a farm where General Delarey was reported to be holding a council of war.

1900
Feb. 12   On the 12th heavy fighting took place. The Boer gun was early silenced, but the outpost line on Worcester Kopjes was attacked and a portion lost. As soon as the enemy's position could be ascertained, THE BATTERY shelled the hill, continuing in action from 11 a.m. to 7 p.m., by which time the enemy was repulsed.

The four companies of the Worcesters lost some sixty killed and wounded, mostly in one company, their Colonel being amongst those slain.

General Clements, who had assumed command on General French leaving for Orange River, now decided that the whole force must concentrate on Rensburg. Orders were therefore given for the outpost line to be evacuated, but THE BATTERY covered the retirement, and reached camp long after dark.

Feb. 13   At midnight the whole force marched for Rensburg, leaving sufficient tents standing to successfully deceive the enemy next morning. An arduous march was concluded at 6 a.m. on the 13th, by which time the whole force, about 4,000 strong, had evacuated the Coleskop-Slingersfontein line and rendezvoused

Feb. 14   at Rensburg. The Left Section rejoined that day. At midnight the force fell back on the strong position of Arundel, bivouacs being again reached at 6 a.m.

Unfortunately two companies of the 2nd Wiltshire Regiment had not withdrawn, and the mounted troops immediately retraced their steps in the hopes of rescuing the party. It was found to be too late, though some twenty men of the Bedford Regiment were met and brought in. After a slight engagement the force returned to camp, which was only reached at noon.

Feb. 15   At 3.30 a.m. on the 15th the Left Section was transferred to Berkshire Kopjes, on the right flank, while later the Centre Section took part in a reconnaissance where slight opposition was met.

Feb. 16   On the 16th, the Centre Section moved out to Dragoon Hill, on the left flank, but returned the same night.

General Clements issued the following order : " The G.O.C. wishes to place on record his appreciation of the spirit and determination of his troops in the operations of the 9th to 14th inst. The powers and endurance of the troops were fully

D

1900 taxed, and they well withstood the strain. The G.O.C. wishes his thanks conveyed to all ranks of the force."

Feb. 18 On Sunday, the 18th, the wounded General Pretorius was handed over to the Boers, and eighty wagons were reported to have arrived at the main laager. It was subsequently ascertained at Colesberg that one contractor there was supplying forage for 10,000 Boer horses.

Feb. 18 The weather was very hot during the day, and the flies and dust extremely trying. Heavy storms on the 19th improved matters somewhat. The nights since the 12th were cool, and even cold.

Feb. 20 On the 20th the following mounted troops proceeded to Berkshire Kopjes under the command of Major Enthoven : 4 guns of THE BATTERY, 1 Squadron Inniskillings, 120 Australian M.R. (principally Victorian M.R.). The enemy with a field gun was already engaged with the garrison of the kopjes, including the Left Section of THE BATTERY. The mounted troops made a wide detour, and after a sharp affair drove the Boers from a range of six kopjes, thus frustrating an attempt they were making to cut the railway. Our casualties were very slight, though at one time the Boer shrapnel were accurately burst. They had not, however, allowed for the strong wind. The enemy left many traces of their losses.

Feb. 22 On the 22nd a move was made to the left flank, Colonel Page-Henderson commanding a mounted force consisting of " J " Battery, 2 Squadrons Inniskilling Dragoons, 300 Mounted Infantry (West Riding, Prince Albert Volunteers, and Australians). He was to meet 400 Mounted Infantry from Hanover Road, near Mooifontein Farm at daybreak.

The enemy was surprised and engaged. A pom-pom on the left flank proved very difficult to locate, but did no damage, although it fired a very large number of rounds. Four guns, with the aid of aiming-posts, shelled the main position, where there was a second gun. In spite of the short range, not a casualty occurred in THE BATTERY.

On the Hanover Road, troops appearing, a flank attack was made on the enemy's right, the Left Section taking a prominent part. The Boers fell back, and by nightfall we had the whole of their position in our possession. It was

1900 strongly entrenched, and prisoners admitted 600 having been present.

Feb. 23 We bivouacked on the ground ; next morning advanced to Elandsfontein Farm, and thence on the range of hills covering Kuilfontein Farm. This position proved to be very strongly held, a 15pr. and two pom-poms engaging THE BATTERY at once. The position was shelled for the whole of the rest of the day, but no material progress was made. At 7 p.m. orders were received to return to Arundel, and the camp was reached, after a very difficult march in the dense darkness, at 10 p.m., only to find that orders to bivouac on the field had missed us.

Feb. 24 At 2.30 a.m. we were again in motion; this time the moon assisted us, and we reached Elandsfontein Farmhouse at 4.45 a.m.—daybreak—on the 24th.

A dash was made through a nek round the enemy's right, and four guns came into action in the open after a four-mile gallop, clearing the nearest kopje in half-a-dozen rounds. Working still further round the flank, THE BATTERY came into action at 2,600 yards, shelling "Spitzkop" from exactly the reverse direction that the Right Section (which had been left) was firing in. The West Riding M.I. attacked from the third side. In an hour and a half this hill was ours, but the next ridge, a very rugged one, offered a determined resistance, and was reinforced by the enemy. About 9 a.m. orders were received for THE BATTERY to fall back to the nek, but the Officer Commanding represented that he was strong enough to maintain his position, and Colonel Henderson sanctioned his remaining. A few rifles with black powder were used by the enemy and were promptly subdued. Such an easy target was not again met with. By this time the greater part of the Arundel force, under General Clements himself, was engaged on the right, including the two 5-in. guns, which succeeded in badly damaging one of the Boer 15-pounders. Meanwhile THE BATTERY had shelled farms on its left flank and rear, as it stood in action, as the enemy were in turn trying to out-flank us. At the latter farm the Boers are known to have suffered loss, and 30 or 40 of them dispersed. Captain Cameron, of the Tasmanian M.I., who was wounded and taken prisoner, afterwards reported that the Boers were confident

1900 that THE BATTERY would be taken. A very long day in action was ended by a terrible storm, and a bad night was spent at Elandsfontein, which was reached with the greatest difficulty. THE BATTERY had no casualties, except horses, but the force had altogether several officers and some 30 men killed and wounded. The Boers admitted 11 killed and 19 wounded—an undoubted understatement, as casualty lists subsequently found in Pretoria contained the names of considerably over 100 burghers.

Feb. 25 During the night the enemy evacuated the position. Colonel Henderson next morning thanked all ranks for the three days' successful work, and later in the day his command returned to Arundel. The boggy state of the ground, owing to the storm, made the march at first very arduous.

Feb. 26 On the 26th Lord Kitchener, who had just arrived, inspected the camp and expressed his approval of the state of THE BATTERY'S horses. Up till this date THE BATTERY had fired 2,500 rounds.

Feb. 27 The advance began next morning, when the Right Section moved out to Jasfontein and surprised a Boer picquet at cards.

Feb. 28 The remaining four guns moved on Rensburg on the 28th with the main body. No serious opposition was made. The

March 4 advance was not resumed till March 4th, when Joubert's Siding

March 5 was reached; but heavy rain delayed the force till the next afternoon, Achtertang being the bivouac.

The railway having been sufficiently repaired, a further

March 8 move was made on the 8th, and Van Zyl's Farm and Siding, near Norvals Pont, reached.

After various reconnaissances, the kopjes on the Free

March 12 State side of the river were bombarded on the 12th. Little response was made, and a considerable force of the enemy was seen entraining at Donkerspoort.

The Right Section had continued, meanwhile, on the right flank and been in action several times.

March 14 On the night of the 14th, the force occupied positions along the river west of the bridge, THE BATTERY entrenching itself

March 15 securely in the dark. In the early morning the pontoon bridge was thrown across the Orange River without a shot being fired.

1900
March 17   THE BATTERY crossed the river on the 17th, Zoetfontein being the rendezvous.

March 19   On the 19th, orders to advance on Phillipolis were at last
March 20   given, but General Clements' command was reduced later to a Brigade, with details, and the rest of the force returned to Norvals Pont.

The following order was issued by General Clements:
"Orders have been received that the following units of the Royal Regiment of Artillery will remain at Norvals Pont and await orders:

> "J" Battery R.H.A.
> 4th Battery R.F.A.
> 8th Battery R.F.A.
> 87th Battery R.F.A.
> 36th Company, S.D., R.G.A.
> Ammunition Column.

The G.O.C. cannot allow these troops to leave his command without thanking all ranks for their cheerful and ready obedience to his orders, and for the great assistance they have rendered to him."

The 8th and 87th Batteries had only just joined the command.

March 30   During the next few days the opportunity was taken to refit, and Potfontein—the scene of the first engagement—was visited, a three and a half hours' ride. It proved a most dangerous position to have attacked in the way ordered.

April 2   On the 2nd of April THE BATTERY marched for Bloemfontein at two hours' notice, being accompanied by 1,300 Mounted Infantry and one Company A.S.C., the whole under the Officer Commanding THE BATTERY. Donkerspoort was
April 3   the limit of the day's march, but on the 3rd of April an arduous march through the pass to Prior's Siding was negotiated.

April 4 & 5   The two following marches were to Springfontein and Jagersfontein Road Station. Here, owing to a rumour of Boers being about, the columns on the march were reconstituted, THE BATTERY joining Colonel Gilbert Hamilton, who had two squadrons of his Regiment (the 14th Hussars), 120 Australian

1900 Mounted Infantry, and 500 Infantry drafts. We were to guard a convoy of 75 ox and 27 mule waggons. Kruger's Siding and Edenburg were the stages on the 6th and 7th, a thunderstorm during the intervening night making the latter day's march a very long one owing to the swampy ground.

April 8 Next day, crossing a tributary of the Riet River, Bethanie was reached, General Gatacre being in command there. The **April 9** march was resumed on the 9th, though an alarm delayed the start. The ox waggons were left at Bethanie. Our bivouac that night was four miles beyond Kaffir River Station, a large force being already on the ground under General Pole-Carew.

April 10 On the 10th, Karl Spruit was reached, and on the 11th, **April 11** Bloemfontein, where we pitched camp near " T " and " Q ' Batteries.

April 14 On the 14th, the 15prs. were exchanged for 12prs., we having then fired 2,814 rounds. Quick-release harness followed a few days later.

We now became brigaded with " M " Battery as Corps Troops, but that Battery left on the 19th for the Mafeking Relief Force and was not again with us. The ammunition column under Major P. E. Gray also joined, and remained with THE BATTERY till all such columns were demobilized late in the year.

April 23 On the 23rd, THE BATTERY marched at a moment's notice to Springfield to support a movement to the east, but on being relieved by the 2nd Battery it returned to Bloemfontein. Later in the war it renewed the acquaintance then made with General Smith-Dorrien.

Heavy rains affected both horses and men ; the former suffered from bilious fever, while both Lieut. Monkhouse and Surgeon-Major Power were sent to hospital with fever. Neither rejoined, Lieut. Monkhouse being shortly invalided home. Lieut. C. H. Wood joined from an Ammunition **May 5** Column on 5th May, and next day THE BATTERY marched **May 6** through Donkershoek and joined the 1st Cavalry Brigade under Colonel Porter, at Nells Drift.

Passing through the Glen the whole bivouaced at Rhode-hoeval—a thirty mile march.

1900
May 7
May 8   Passing through Brandfort on the 7th, THE BRIGADE halted at Eensgevoden, another march of nearly thirty miles. The Wet River was crossed early next morning, and shortly afterwards the column marched past Lord Roberts, who expressed his interest in a Battery from India, and his approval of the state of our horses : Smalldeel station was the bivouac.

May 9   On the 9th THE BATTERY marched with the 11th Division to Welgelegen.

May 10   In the advance on the Zand River next day THE BATTERY moved close to Head Quarters and Major Enthoven received personal orders from the Commander-in-Chief to join the 3rd Cavalry Brigade (Brig.-General Gordon) and assist it in the passage of the river. The Battery was not in action till the river had been crossed, when various positions were taken up to press the retreating Boers. With the exception of a pom-pom late in the day, the enemy's guns made but indifferent practice.

The Brigade pushed on till dark, and halted at Ventersburg Road Station.

May 11   A flock of sheep had to provide rations on the 11th, and THE BATTERY went into action with the carcases slung under the axletrees.

The enemy was found entrenched at Boschrand, nine miles from Kroonstadt, and opened with two field guns (one a 12pr. B.L.) and a pom-pom. The Left Section, assisting Colonel Henry's Mounted Infantry on the left flank, silenced the 12pr. in three rounds. The second field gun also withdrew early, but the pom-pom opened at intervals throughout the day, till the flash showing up in the dusk disclosed its exact position. The trenches were also thoroughly searched. During the night

May 12   the enemy hastily withdrew, and on 12th May Kroonstadt was entered. THE BATTERY headed the Army, which now halted for a week.

150 miles had been marched in six days, and two actions fought ; rations at first were scanty, groceries being very scarce.

During our halt here we heard of wounded Boers having been brought in from Boschrand as the result of THE BATTERY's fire, and it was reported that the gun teams had also suffered severely.

1900
May 17
General French inspected the horses on the 17th, and expressed his satisfaction. One hundred and nineteen Walers were still left of the 198 embarked, while 12 had been transferred to " M " Battery at Bloemfontein.

A metal-lined case was picked up on the march containing cartridges; it was addressed to Commissariat, Pretoria, via Lourenço Marques, so had apparently entered as food. It was re-addressed to General Grobbelar.

May 21
The advance recommenced on the 21st, when a new corps was formed under Colonel St. G. Henry. It consisted of

THE BATTERY,
Ross' Colt Gun Battery,
4th M.I. Regiment,
8th M.I. Regiment,
Two Companies South Australian M.I.,
One Company Victorian M.R.,
One Company Tasmanian M.I.,
Two Companies 7th Imperial Yeomanry,

and THE BATTERY remained with this force till its return to Pretoria from Komati Poort in October.

Halting that night near Jordans Siding, Henry's command became the screen for the 11th Division and Head Quarters of the Army.

May 22
On the 22nd Wolvelaagte, near Serfontein Station, was the bivouac. The country hereabouts was very open rolling veldt a great deal of which had been burnt. Dispositions were made

May 23
for an engagement next day at the Rhenoster River, but, though very carefully entrenched, it was not held, and the crossing was made without difficulty except that offered by very steep and high banks.

Near Leeuw Spruit Station traces of a very large laager were found, and a splendid dam being at hand the force halted·

Boer newspapers were found giving an account of our " heavy losses " at Boschrand ! Later on their papers spoke of the " vigorous defence " of the Rhenoster !

Forage was scarce, and but for some oat-straw found hidden in a farm the horses would have fared miserably.

On the 24th Vredeford Station was reached at 9, and crossing the railway the march ended at Kromellanberg Spruit, where General Ian Hamilton's Division came up and bivouacked on our right.

In honour of Her Majesty's birthday, other troops drank the Queen's Health in rum, but that for THE BATTERY apparently remained with the Corps Troops elsewhere. It was some weeks before our rations were entirely separated from Head Quarter supplies.

May 25 The bivouac on the 25th was at Rosebery Farm, a somewhat swampy plain, and next day the Vaal was crossed. Starting early and moving through cultivation, Viljeon's Drift, was found to be held. Crossing a deep mullah with some difficulty, THE BATTERY came into action at 3,000 yards, and the Boers fled at once. The Right Section then engaged the enemy, holding the big coal mine across the river. The Mounted Infantry were shortly enabled to cross by the drift, and THE BATTERY following, shelled Vereeniging station heavily.

The Boers retreated so rapidly that only one span of the railway bridge was destroyed.

The part owner of the mine presented THE BATTERY with a waggon and team of mules for its assistance in saving his mine, but later on claimed more than their full value from Government!

Though the enemy had cleared, THE BATTERY was brought back to the south side of the river at dusk. Re-crossing next **May 27** morning, Colonel Henry was suddenly ordered at mid-day to rescue the 3rd Cavalry Brigade, supposed to be held at Englebrecht's Drift to the East. A very arduous march over sand and drifts resulted in nothing being found but a few of the enemy's scouts, both the Cavalry and Boers having retired, the former on Vereeniging by the south of the River.

Bivouacking very late at a Kaffir kraal on the Vaal near **May 28** Panfontein, the force was in motion again at 4 a.m., a hard frost increasing the difficulties.

Crossing the Zuikerbosch and the Klip Rivers we made our way through the 11th Division, passing from the right rear to five miles in advance, and halting, after a thirty-five mile march, near Klip River Station. This double forced march

cost us six horses abandoned and some ten more unfit for
further work.

Some entrenchments were found near the station, and a
large number of entrenching tools were lying about; the
trenches had not been occupied.

Frost set in again at nightfall, and by morning buckets
showed nearly half-an-inch of ice.

Pushing on on the 29th as fast as the state of the horses
would allow, touch was obtained of the enemy near Nell Junction,
where the first of the Johannesburg mines met with is situated.

A pom-pom opened fire on both the Mounted Infanty and
THE BATTERY, and Colonel Henry ordered THE BATTERY into
action. This message was conveyed by Captain the Duke of
Norfolk, who was galloper for the day.

Crossing the railway near Rooikop Station to gain a well-
situated kopje, an 18pr. also opened, but the percussion shell,
though well directed, did no damage.

This gun was located and a heavy fire directed on it. The
pom-pom gave no further trouble, but a 12pr. (captured) gun also
engaged us from a nearer kopje. The time shrapnel was burst
too short, and no material damage was done; but the fire was hot.

After some little time its position was discovered and a few
rounds soon silenced it. Captain de Villiers, a Frenchman—
the Staats Artillery Officer in charge—was taken prisoner a
few days later, and informed Colonel Henry that five of his
detachment were killed and the gun disabled.

The 18pr. shortly afterwards found it advisable to withdraw, and THE BATTERY pushed on to the Centre Section,
previously sent forward with the most advanced portion of the
Mounted Infantry.

THE BATTERY was next brought into action on a hill overlooking Elandsfontein. A 98pr. (on a truck) was shelling this
hill heavily, and aiming posts had to be used. The Pretoria
line was the objective, and the first round caused a stationary
train to move and another following to run back. This and
six other trains were subsequently captured.

The sight of the guns also caused the suburb of Boksburg
to send a flag of truce to Colonel Henry. The fighting
burghers shortly withdrew, and the 98pr. steamed away just

1900 before Colonel Ross and his Mounted Infantry reached the railway north of the junction and cut it.

THE BATTERY moved on to a hill close to Germiston and halted, the horses being done up after the long and rapid advances.

Lord Kitchener came up to see the captured engines and rolling stock, and told THE BATTERY to bivouac where they pleased and take what supplies there were in the trucks. Unfortunately they were useless.

All rations and forage had now run out, but some chickens and mealies were finally found. We bivouacked in some empty houses.

An Australian newspaper correspondent who was with THE BATTERY sent a most flattering account of its behaviour to his journal, and on June 1st Colonel Henry issued the following order :—

" It is published for information that the Commander-in-Chief has expressed his great satisfaction at the work done by the force under Colonel Henry's command in the operations on the 29th ultimo, which resulted in the capture of seven engines and a large amount of rolling stock at Elandsfontein Junction. Colonel Henry wishes to thank the Officers, Non-Commissioned Officers, and men for the manner in which they assisted in obtaining the object of the Commander-in-Chief."

May 30 THE BATTERY moved next day to the north side of the mines, where the Corps formed a camp. Three ammunition wagons were left at the railway station, as there were now insufficient horses to draw them.

May 31 On Thursday, May 31st, orders were suddenly received to move to the water-works five miles away. Arriving there Colonel Henry and his staff, with a strong escort, proceeded to the Johannesburg Fort—a mile distant—and took it over from Commandant du Preeze. THE BATTERY'S 1st Trumpeter sounded the general salute when the Union Jack was hoisted.

In the Fort were found an English 6·2 howitzer (over 20 years old), four 7pr. guns (one badly damaged at Mafeking), two machine guns much injured by shrapnel, a large amount of ammunition, and some 2,000 rifles, including all those taken in the Jameson Raid.

Later in the day THE BATTERY moved four miles to the north, the bivouac being the far side of the woods, and conveniently near to some market gardens.

June 3
The advance was continued on the 3rd of June, and camp formed at Deep Sluit Farm.

June 4
Marching at 6.30 next morning, a section was soon pushed forward to assist in the passage of the Hennops River (Zes Myl Spruit), and a few rounds cleared the way for the 7th Yeomanry. At the top of the next ridge they were again checked by rifle fire, and THE BATTERY was again ordered up. The Southern Pretoria Forts were then in sight, some 8,500 yards off, but two 18prs. and a pom-pom opened a very hot fire from an intervening ridge. These guns were well concealed and protected, and before they were silenced several slight casualties occurred. The range was 3,700 yards, but fortunately the Boers were principally using percussion segment shell. One pitched absolutely under the muzzle of No. 1 gun, but no one was hurt by it. Word, however, reached Head Quarters that THE BATTERY was suffering severely, and an order was received to run no risks, as the heavy guns were coming up. When the Staff Officer arrived with his message THE BATTERY was standing fast, having temporarily silenced the enemy's fire, both gun and rifle.

This was one of the rare occasions on which Boers were freely seen; they showed up on the sky-line, and the kopjes were searched by our shrapnel.

The position selected for THE BATTERY being 100 yards on the reverse slope accounted partly for the slight casualties.

An hour and a-half after opening fire in this position, two 4·7 naval guns came up and the section on the right was withdrawn, upon which the enemy's pom-pom again opened fire. An accurate sketch of this scene appeared in the *Graphic*.

Leaving the Right Section in action, the rest of THE BATTERY moved to the extreme left flank, but was not again engaged, though under fire.

The action lasted till darkness set in, when there was great difficulty in collecting Colonel Henry's Corps and obtaining water and supplies.

1900
June 5
On the 5th June Pretoria surrendered, and THE BATTERY, entering among the first, camped about a mile north of the racecourse. Of the 170 horses that left Bloemfontein with the Battery, 85 were really fit for work, 43 more were brought in, and 42 were deficient.   Of the 115 Walers, 71 reached Pretoria.

June 7
On the 7th orders were unexpectedly received at mid-day to march to Koedoespoort. Only two Sections were taken, the Right remaining in Pretoria with the crippled horses.

June 8
Next day, passing through the Cavalry and 11th Divisions at Hartebeast Drift, Struben's Drift (near Marks' Farm) was reached. When about to bivouac a 98pr. fired right over THE BATTERY, making good practice on the track just quitted. Recrossing the drift a camp was selected some distance further back, while the 5in. R.G.A. guns engaged the " Long Tom " with great accuracy.

June 10
On Sunday, the 10th, the Queen's and other telegrams, and Lord Roberts' Army Order on the Bloemfontein-Pretoria march were read out at Church parade.

Captain Corbyn was sent into Pretoria sick, but was not struck off the strength till the 22nd. An accident to his foot, followed by jaundice, was the cause of his leaving.

The negotiations believed to be in progress coming to nothing, preparations were made late that night for the next two days' engagement, now known as " Diamond Hill."

June 11
The 8th M.I. Regiment having been left at Pretoria, Colonel Henry had but 700 men and THE BATTERY. With this force he had to hold three and a half miles of front, from Marks' Farm to the north, and the range of hills starting 6,000 yards from the main Boer position and running east and west.

Shots were exchanged with the Boer outposts from the eastern extremity of the hill, near Bavaans Poort, and patrols to the north of it were engaged. The general action on both flanks could be seen, and the 98pr. in Pienaars Poort, but the latter did not fire.

Unless a counter attack was made by the enemy this part of the line could not be seriously engaged. At dusk THE
June 12
BATTERY returned to its former camp, and on the 12th repeated the previous day's experiences. Two Boer guns opposing the

1900 Cavalry division were seen in action, but were entirely out of range, being over 6,000 yards from the nearest point THE BATTERY could approach.

The bivouac that night was south-west of Bavaans Poort.

June 13 Hearing next day from a deserter that the Boers had retreated, THE BATTERY followed up the reconnaissance at once; but camp was formed that night east of Bavaans River, not far from the last bivouac.

June 14 On the 14th Colonel Henry's force advanced to Pienaars
June 16 Poort. Two days later the rest of the troops, except General French's command at Kameel Drift, withdrew to Pretoria. This left a front of 15 miles to be guarded by 500 men and THE BATTERY, the 4th Mounted Infantry Corps having been reduced to that number by the departure of the 7th Imperial Yeomanry.

Placing the Tasmanian M.I. on the left and the Victorian M.R. at the Poort, Colonel Henry took THE BATTERY and 4th M.I. Regiment to Donkerhoek, the South Australians being on the extreme right at Botha's Farm. It was not believed that the Boers would return after having been pursued to Bronkhorst Spruit; but it was soon evident that this idea was
June 17 erroneous. Signs of the enemy were seen on the 17th, and on
& 19 the 19th shots were exchanged all along the line of outposts. THE BATTERY was in a bad position, as it could not come into action without falling back, the high ground to its front being inaccessible from the west. The reappearance of the enemy caused the 11th Division to return to Marks'
June 21 Farm on the 21st, and THE BATTERY joined the main body there.

June 24 On the 24th, after a frosty night, camp was shifted to a
June 26 position between Marks' Farm and Eerste Fabricken, but two days later THE BATTERY was hurried once more to Donkerhoek,
June 27 moving the next morning, the 27th, to Botha's Farm, where the South Australians were still posted. Above the farm is the plateau, five miles in circumference, on which was fought the final stage of the Diamond Hill engagement. An outpost company of the Guards now held it.

June 28 There was a great deal of sniping on the 28th, so a Section of THE BATTERY was taken by the Commanding Officer on to

1900 the plateau, surprising the Boers and inflicting sufficient damage to restrain them from reappearing for ten days.

The same evening THE BATTERY returned to Donkerhoek, bivouacking on the south side of the stream next to the 3rd Grenadier Guards. This was the 10th change of bivouac in twenty days, all within a radius of about five miles.

One Section went out to Botha's Farm—two miles—daily, but were not in action till July 7th. A Boer who surrendered **July 1** on July 1st stated that he saw some men and horses killed by our shrapnel on the 28th. He also stated that Erasmus, Viljoen, and Delarey (in the order named) were in front of us; Botha at Balmoral, and a total strength of 6,000 men. Also that there was abundant ammunition except for the Creuzot guns.

July 2 & 4 On the 2nd and 4th all four guns were sent out at dawn to the farm, but the alarm proved unfounded. Nothing was seen.

July 6 The Right Section rejoined from Pretoria on the 6th with twenty-five remounts, and most of the horses left with them a month before to recover.

July 7 On the 7th the Left Section returned at noon from the outpost. Two hours later news was brought that the South Australians were being attacked. The Section re-harnessed, hooked in, and reached Botha's Farm in twenty-three minutes, only to find that the Boers—about one hundred—had exchanged shots at eighteen hundred yards and retired.

Some of the enemy had also sniped the Guards on the plateau, the first time since June 28th. Creeping up, the Section opened with time-shrapnel, the range and fuse having been recorded on the former date. The snipers retreated precipitately, and never returned again.

July 8 On Sunday, the 8th, the whole BATTERY paraded for Divine Service, as the Boers had so far respected the Sabbath. But there was a repetition of the previous day's work, some 140 men—Zarps, apparently—approaching Botha's Farm. They did not, however, come within range.

July 9 The following day, General Hutton being attacked at Reit Vlei, Colonel Henry's force was sent out to support him. The whole of THE BATTERY accompanied it, and moved some six miles to the south towards Witfontein. Touch was obtained

1900 of the Boers retiring and helio-communication established with General Hutton. As the enemy had been repulsed and it was now late the column returned to Donkerhoek.

July 11 At 6.30 a.m. on the 11th Colonel Henry again moved out, taking four guns with him, the Right Section being sent on to the plateau, from whence it silenced a gun that opened on us as we moved off to the south.

Touch was again obtained with General Hutton, and a few rounds expended at Boer picquets, which hastily retired. Another Boer gun attempted to shell THE BATTERY, but was quite unable to reach it. Orders were at that moment received to return forthwith, and marching straight through our camp, the whole pushed on to Eerste Fabricken. Here horses were watered and fed, after twelve hours' fast. THE BATTERY (four guns) then pushed on to the Bavaans Poort range of hills, which was to be held against the north. The reverses at Uitvals Nek was the direct cause of this change of front. Vlakfontein was close by, the bivouac of the 12th June.

The following day was spent in preparing emplacements on the hills and making roadways up to them. General Pole-Carew inspected the position. THE BATTERY was now hooked in daily by dawn and the coldest hour of the 24 was thus spent, waiting and watching the water freeze !

July 15 On the 15th, two pom-poms, under Captain G. T. Brierley, joined THE BATTERY, also a few remounts.

July 16 The Cavalry Division worked across our front on the 16th and two Boer guns and a pom-pom opened on them at extreme range. Our 5in. guns near Eerste Fabricken also shelled the northern side of Pienaars Poort, from whence a West Australian picquet had been driven.

Henry's Mounted Infantry was ordered up to reinforce, and at once moved off with the pom-pom section, leaving THE BATTERY and 90 South Australians to hold the Bavaan Hills.

July 17 A company of the Buffs arrived during the night, relieved the picquets in the morning, and THE BATTERY rejoined Colonel Henry at Pienaars Poort. It reached the Donkerhoek Camp later in the day, and found that the Right Section had been in action during its absence and fired a few rounds.

1900
July 20
Captain D. Arbuthnot, posted *vice* Captain Corbyn, joined on the 20th.

July 21
The following day the operations leading to the occupation of Komati Poort may be said to have commenced. Colonel Henry's troops moved to the south at daybreak, the objective being a hill known as Morskop. Beyond this hill again was Witfontein. The Mounted Infantry moved round the southern end of the former kopje, and were soon engaged with the enemy. The Pom-pom Section followed and came into action, while THE BATTERY vigorously shelled the hill from the west. The Boers were reinforced, and this being simply a reconnaissance, the Mounted Infantry retired at leisure; the shrapnel almost silencing the Mausers.

Two guns now opened on us from Witfontein, but found the range altogether beyond their power. Advancing on to Morskop, their smaller gun—apparently a 9pr.—was still unable to reach us. Not so the 15pr. THE BATTERY had anticipated this move, and that the Boers would come into action beyond the range of its 12prs. The Officer Commanding, therefore, requested permission to advance a mile, but was ordered to remain where he stood, as instructions were momentarily expected from Head Quarters.

The larger gun dropped shell and bullets round us, but turned their attention principally to the teams three hundred yards in the rear. Several horses were hit, but the shell all being burst too short they were little the worse.

The range was well over 5,000 yards, still by sinking our trails sufficient elevation was obtained, though the shooting was erratic. In twenty minutes the Boer guns were silenced, and passing the spot two days later, a pipe covered with blood was found, and at a farm not far off, the owner's grave.

On the way back another 15pr. and a pom-pom showed up, but did no damage. This force was the commando retiring from before General Hutton, and it was subsequently ascertained that they lost three killed and several wounded; our loss was nil.

July 23
Orders and counter-orders on the 23rd resulted in THE BATTERY standing in marching order for five hours and finally only moving to Botha's Farm.

E

1900
July 24   The following day the advance began, and, firing but a single round, THE BATTERY moved past Witfontein. Colonel Henry's corps formed the advanced and right flank guard to the 11th Division. Crossing Bronkhorst Spruit and the field of battle, the graves of those of the 94th Regiment who fell in '80 were passed. Bivouac was formed half a mile further east at one of the numerous Nooitgedachts—a twenty-seven mile march as the Battery moved.

July 25   On the 25th THE BATTERY worked to the south and camped at Witpoort, on the Wilge River. Soon after picketing, a storm broke, and lasted through the night. Captain Arbuthnot went back to Pretoria that morning, sick.

July 26   The ground was so water-logged that only a very short move was made late next day to Klipfontein.

July 27   On the 27th Brugspruit was reached, the camp being on the railway three-quarters of a mile from the station. Water was here a difficulty, and the weather cold at nights.

July 29   A reconnaissance 15 miles to the north was made on the 29th, THE BATTERY supporting it. Touch was obtained of the Boers in the bush veldt.

July 31   Nooitgedacht Colliery was situated three miles to the north, and on the 31st Colonel Henry's Corps moved there. A well-placed camp, but the water disagreed with the horses. A Field-Cornet Erasmus and 13 men surrendered next day.

Aug. 4   The advance continued on the 4th of August, Olivants River being reached. The bivouac was south of the iron bridge, one of the very few road bridges in the Transvaal. The veldt was now very dry, and several times caught fire ; one of the Ammunition Column's waggons was partly burnt, but the contents escaped. Middleburg, 95 miles from Pretoria, was entered next day. The Centre Section was sent to Uit Kyk, a station five miles to the south-west, where two Companies of the Suffolk Regiment were strongly entrenched.

Aug. 7   On the 7th THE BATTERY shifted from the west to the east of the town, and the Centre Section rejoined. Cold winds had now set in, and caused much discomfort at times ; frost was frequent at nights.

**1900
Aug. 11**   Captain Arbuthnot returned to duty on the 11th, and ten remounts were received, six being cavalry horses with sore backs, but fit to work in draught on the off side.

Aug. 19   After a fortnight's halt the advance was continued on the 19th. Crossing the Klein Olifants River, Pan Station was reached, THE BATTERY bivouacing a mile south.

Aug. 20   The following was a hot day, and no water obtainable for 16 miles. Even on camping near Wonderfontein Station only water oozing from the ground was to be found.

Aug. 21   One pom-pom went out two miles on the 21st and dispersed a Boer patrol. In the afternoon Henry's Corps moved to Leeuwfontein, a few miles to the south, where there was both water and grazing.

Aug. 24   Belfast was attacked on the 24th. Moving north of the railway we turned to the north-west at Paardeplaats, and drove the enemy's scouts off Klipfontein Hill. Reinforced, these men checked our advance guard on the next ridge, but one shell at 4,000 yards cleared them out. This, a lucky shot by the Commanding Officer, pitched in the centre of a triangle formed by three groups of two men. One of these men was captured at Komati Poort, and stated that the shell almost touched his head.

Crossing with difficulty the boggy stream running north from Belfast, THE BATTERY came into action at Tweedfontein ; but the enemy cleared at once, and an attempt was made to find a position near the Monument. This stands on Steynplaats three miles north of Belfast Station, and marks the highest point in the Transvaal, over 6,000 feet. But the Boers were so strongly posted at less than 800 yards that it was impossible to get into action here. THE BATTERY was halted half a mile back, out of view, but two of the bullets directed at the Officer Commanding when reconnoitring travelled on and struck Lieutenant Hine-Haycock's horse. Some other horses were also hit. The 18th Infantry Brigade now came up and a pom-pom opened fire. In trying to locate it the same two officers received the particular and almost accurate attention of its shells.

Orders were then received for the 4th M.I. Corps to press round the enemy's right. THE BATTERY accordingly shifted,

1900 but before the Mounted Infantry could secure a position darkness set in. The Corps was fortunate enough to have but few casualties—one officer, six men, and some 20 horses wounded. THE BATTERY bivouaced three-quarters of a mile from the Monument, and the coldest night of the year was passed. A very keen wind was combined with 13 degrees of frost, lack of firewood, and no transport.

Aug. 25 THE BATTERY harnessed up at daybreak, and was about to move when two shells, obviously fired at random, pitched one 30 yards over and one 10 yards short of the bivouac. Being " percussion " fired at extreme range no harm resulted. The Boer 98pr. fired at great range at Lord Roberts' train as it arrived at Belfast, but failed to reach it.

Though supporting a strong patrol that worked north along the Lydenburg Road, THE BATTERY was not in action this day. A horse was, however, wounded in camp by a spent bullet. General Marshall rode round the camp in the evening, and complimented THE BATTERY on its appearance.

Captain Arbuthnot commanded THE BATTERY on the 26th, 27th, and 28th, the Commanding Officer being on the sick list.

Aug. 26 The 18th Brigade pushed forward, and THE BATTERY, late in the afternoon, came into action within seven hundred yards of the Boers, but behind a ridge, most of the bullets appearing to pass over the guns. Only one man and three horses were hit—one of the horses being again Lieutenant Hine-Haycock's. One hundred and twenty-four rounds were fired, and, from the signs found next day on the enemy's position, many were effective.

THE BATTERY bivouacked some two miles from its previous camp.

Aug. 27 THE BATTERY, though not in action on the 27th, was under fire from the 98pr., and two men of the escort were severely wounded by it.

Lakenvlei, the bivouac, was again but a very short distance on.

Aug. 28 On the 28th THE BATTERY was with the left flank guard, but not engaged. It camped close to its former bivouac.

Aug. 29 Vinaars Poort, a rugged pass, was crossed next morning. It had been extensively prepared for defence with strong em-

1900 placements for big guns, but was not held. The next few miles were very mountainous, and the descent from the pass necessitated drag ropes, as well as brakes.

Another rough pass was climbed in a thick mist, which lifted later, and showed Helvetia to the S.S.E., and Machadodorp to the south.

Descending on to Vlakfontein plain the 4th Corps formed the left flank and bivouaced at dusk at the farm of that name.

Boers were seen to the north, but kept clear of the Mounted Infantry picquets. This was the end of their boasted stand on the Machadodorp Mountains.

Aug. 30 Colonel Henry pushed on rapidly through Helvetia on the 30th to Doornhoek, above Waterval Onder, and relieved the Cavalry Brigade there. The camp stood nearly two thousand feet above the railway station and village, and a splendid view of the Elands River Valley was disclosed. Snipers were firing on Waterval Onder held by our Cavalry, and a Section of " T " Battery was firing across the valley at them. This was relieved by " J " Battery's Centre Section, but it was impossible, with the vast hill opposite, to locate the enemy.

The released prisoners from Nooitgedacht began filing past the camp at 5.30 p.m., and the string continued at intervals till 8.30 a.m. next day.

Aug. 31 On the 31st the Right Section replaced the Centre Section to watch the valley, and the Left, taking up a commanding position on the western edge of the plateau, covered Waterval Boven. The former only fired, and but a few rounds.

Sept. 1 Two Sections were again in position on 1st September, but were not engaged, and next day, Sunday, things were sufficiently settled for the railway town of Waterval Boven to be visited.

Sept. 6 An expedition to the south of the valley was commenced on the 6th, with the double object of assisting General French's advance on Barberton and clearing the Elands Valley. To reach the far side of the valley it was necessary to return as far as Machadodorp—some 13 or 14 miles. The road down from Helvetia was in a dreadful state, steep, thick with dust, and covered with dead animals, while an empty convoy from Lydenburg blocked the way. The march occupied over six hours.

The following day the force (which consisted merely of 4th M.I. Corps) toiled up the long steep hill to Nooitgedacht, yet another farm of that name. A few Boers at the top fired a volley and fled. Their laager was located about four miles to the south, and the Right Section was taken rapidly on to shell it, while the South Australians worked round to cut them off from the Komati River, some eight miles on. But the commando had received the alarm, and was off at once. A running fight took place, the enemy, using a pom-pom on the north side and a 15pr. from the further side of the river, finally disappearing over the high hills to the south. We bivouaced on the above-named farm due south of Waterval Boven. Our casualties were immaterial.

A thick mist throughout the whole of the next day made any movement impossible, and it was not till 8 o'clock on the 9th that operations could be continued. The Boers were found strongly posted at Weltevreden a few miles further to the east and within sight of Doornhoek.

THE BATTERY, one pom-pom, 4th M.I., and Tasmanian M.I. made the frontal attack, the Victorian M.R. and one pom-pom the flank attack on the enemy's left, shortly reinforced by the Centre Section. THE BATTERY came into action at 3,000 yards on the enemy's centre, and before long subdued his fire and forced him to retire; he had no guns. THE BATTERY pressed on and closely followed the Mounted Infantry on to the position it had shelled. It was now possible to enfilade the Irish-American Brigade, which formed the Boer left, and they fled in disorder into the valleys behind.

Twenty horses were captured, Captain Brierley driving one group in by dropping pom-pom shells behind them.

General Hutton arrived as THE BATTERY first came into action, and a part of his force shortly after came up, consisting of Brabant's Horse with their two guns. They attacked the Boer right flank posted on a long and high hill, and obtained a footing on it. The Left Section, therefore, moved somewhat to the north and co-operated. This manœuvre once more enfiladed the enemy at about 4,000 yards, and they hastily withdrew. Five dead and several wounded Boers were found. One,

1900 an American named Luther, stated that they had 500 men present under Viljoen, and that their centre fell back so rapidly that the left did not know they had gone. He was brought in in a Dhanjiboy ambulance with his friend O'Hara, both shot by shrapnel. The latter was dead when he reached us, and the gunners buried him.

Our loss was practically nothing, though the position was admirably adapted for defence. The rout was so complete that no further opposition was encountered in the mountainous country between this and Kaap Muiden.

In connection with this engagement, it is worthy of note that THE BATTERY took a prominent part in the Battle of Welterwreeden (a mere variation in the spelling of Weltevreden) in Java just 89 years before, viz., 10th August, 1811 (*vide* p. 8).

We bivouacked on the ground we had won, and General Hutton paraded THE BATTERY to compliment it on its dash and excellent shooting.

The rest of his force came up next day, consisting of the 1st M.I., a section of pom-poms, and other details. The Brabant guns were a worn out Hotchkiss—the shell no longer taking the rifling—and a 15pr. of the 77th Battery recaptured from the Boers, also much worn, but serviceable.

Sept. 10 On the 10th a mountainous march brought THE BATTERY close to Uit Komst, one valley into which we dipped requiring ten horses in each team to surmount the following rise. A handful of Boers held one large kopje and Colonel Alderson was checked. THE BATTERY (Right and Centre Sections) soon shelled them off, and the whole force camped a few miles on above the huge Godwan Valley, beyond which Kaapsche Hoop and Tafel Kop tower.

Sept. 11 On the 11th the Centre Section accompanied the 1st M.I. to Hooitbosch above the Elands Valley, Nooitgedacht (the prisoners' camp) being 5,000 yards distant, but far below us. Men were seen approaching the station, and the Officer Commanding was ordered to open fire. Fortunately he made them out to be West Australians, the mounted troops with the Guards' Brigade, and refrained. A Staff Officer climbed down, found that they were so, and that the Guards were still

1900   at Waterval Onder. He took five hours to return up the mountains.

In the afternoon Major Enthoven was sent with a patrol down into the Godwan Valley to see if the track was passable for guns and waggons.

This track descends some 2,000 feet, and is only used by the Boers when moving from summer to winter grazing and the reverse. Ox waggons take twelve hours to come up the three miles. Going down the fore wheels of the wagons are both skidded, for if the hind wheels are stopped the vehicle is apt to slide over the edge of the road unless held from behind. Wrecks of several waggons were visible. The track in one place is for a short distance at an angle of 45°, at other places merely slabs of rock shelving outwards. The ascent to Kaapsche Hoop was said to be similar. It was therefore reported that guns would travel down safely, but would require double teams to ascend the further side of the valley. Wagons would delay the force ; mule-waggons break down, and probably none reach Kaapsche Hoop even with double teams, which were not available. General Hutton therefore decided to advance with two guns (no waggons), of THE BATTERY (Left Section), one pom-pom " J " Battery, one pom-pom Hutton's, most of the Mounted Infantry, but no transport, the guns and pom-poms with double teams. The other four guns of THE BATTERY, two guns Brabants', two remaining pom-poms, all transport and details Mounted Infantry were placed under Major Enthoven, with orders to rejoin *via* Machadodorp as quickly as possible. This latter force amounted to 16 officers, 546 men, 463 horses, 882 mules, 6 guns, 2 pom-poms, 8 ammunition waggons, 126 transport waggons, and 235 natives.

Colonel Henry, reconnoitring to the south on the 11th, came in touch with some Boers, supposed to be cattle guards, **Sept. 12** and it was necessary to take all precautions when the long column commenced retracing its steps at noon on the 12th.

In crossing the deep valley mentioned on the 10th, double teams had to be used, and all the transport was not in camp till 2 a.m. The rear had to wait till the moon had risen.

1900   The bivouac was at Weltevreden, on the high ground the Boer right had defended.

Sept. 13   The South Australians were still posted here, and next day the West Yorkshire Regiment was passed on the way up to relieve them, and also a post of Brabant's Horse at Uit Komst.

Sept. 14   The march was to Nooitgedacht, our camp on the 7th. At Machadodorp, on the 14th, the loads were lightened and, Head Quarters being here, many re-arrangements were made. The

Sept. 15 following day the column marched at 2 a.m. by moonlight, and the top of the obnoxious hill to Helvetia was reached as day broke. Passing on through our previous camp at Doornhoek, the precipitous road down to Waterval Onder was safely passed and bivouac formed on the banks of the Elands River by noon.

The heat in the valley was tropical, and continued so till high ground was again reached a month later. On this

Sept. 16 account the march on the 16th commenced at 4 a.m., the red dust at first being blinding. Near Nooitgedacht the valley opens out and the road becomes fairly level. Watering close to where the prisoners had been confined the force pushed on to Godwan station. Here the column broke up, each detail rejoining its unit, while the Left Section was found in camp close by.

Since leaving THE BATTERY at Uit Komst this section had done some stiff marching. It had descended to the Godwan River without accident, though a few Cape carts that accompanied the force had capsized. The following day the Kaapsche Hoop range was climbed, an almost inaccessible mountain in places. The guns reached the summit before being overtaken by the Sappers sent to assist them, and General Hutton personally complimented the Section on its work. Fortunately there was no opposition. Supplies had now almost run out, so, leaving what were left with 300 men under Colonel Henry, General Hutton and the whole of the rest of the force descended to Godwan ; and here we found them.

Orders were received during the day for THE BATTERY and all details of 4th M.I. Corps to rejoin Colonel Henry and next morning at 5.30 the force started, strength as follows :—

	Men.	Horses.	Mules.
"J" Battery	100	 110	 65
Pom-pom Section ...	18	 22	 13
4th M.I.	38	 28	 170
South Australians ...	200	 170	 52
Victorian M.R. ...	26	 25	 120
Tasmanian M.I. ...	75	 66	 32
Natives	50	... ... —	... ... —
Total	507	 421	 452

After crossing the Godwan River there was a continuous ascent for nine miles, very steep in places. Fortunately a cold mist eased the work for the teams, and the stiff pull was successfully accomplished by 8 a.m. Dropping down a short steep hill into Kaapsche Village THE BATTERY bivouacked on the ground which the Left Section had occupied four days previously.

Sept. 18 General Hutton's force came in later in the day. On the 18th only a short march, to Kaap Vallei, was possible, as some of the transport failed to reached the top of the hill the previous night.

Sept. 19 It rained that night. The descent into the Kaap Valley was again rapid, and the road most greasy. A steep sided drift over the North Kaap River gave trouble. Bivouac was formed at Murray's store, twelve miles from Barberton.

Sept. 20 Turning to the east on the 20th, the South Kaap River was reached, and, keeping mostly along the left bank, camp was formed at the junction of the two Kaap Rivers. The march was but a short one, as General Ian Hamilton's troops were caught up and there was no possibility of passing them. The valley was narrow, hot, and dusty, and shut in by "jungle" covered hills.

Sept. 21 On the 21st Avoca was passed, and at the bivouac, three miles beyond, the following ammunition was found in a tributary of the Kaap :—

6,000 Martini cartridges,
2,000 Maxim cartridges,
2,000 Revolver cartridges.

1900
Sept. 22 On the 22nd our road was again blocked, and Kaap Muiden Station on the main line only reached after a nine hour march. Large trees were the feature of the road. General Hamilton was here caught up, and arrangements were made to start

Sept. 23 before his force next day, but the non-arrival of supplies by train delayed our departure till daybreak on the 24th—a loss of 24 hours.

Sept. 24 The Left Section and Ammunition Column remained at Kaap Muiden. Moving through a deserted jungly country, camp was formed at Malelane.

Sept. 25 On the 25th a double march was undertaken. Starting at 4.45 a.m., Hector Spruit was reached, and a halt made during the heat of the day. In the river close by were the fourteen destroyed guns left by the Boers.

Marched again at 7.45 p.m. There is no water between Hector Spruit and Komati Poort, and this alone would have made night marching in the heat necessary. At 11 o'clock General Hamilton's rear guard, that had started from Malelane 24 hours before, was caught up. Off-saddling for three hours,

Sept. 26 we were again in touch with the troops in front of us at 6. On entering the Lebombo Flats, we were at last able to pass the force that had blocked our way for six days, and arrived at Komati Poort at 8.30 a.m., 24 hours after the 4th Mounted Infantry and Guards had marched in. Colonel Henry was camped at the junction of the Komati and Crocodile Rivers. The heat was great, and a hot wind added to it.

The mass and mixture of engines and rolling stock collected here was marvellous. Six pounds of musty mealies per horse

Sept. 27 was all the forage obtainable that day, but next morning good supplies came through from Lourenço Marques.

The following telegram was received from Lord Roberts and issued by Colonel Henry in his orders on the 27th: " General Pole Carew, accept yourself and convey to General Inigo Jones and all concerned my congratulations on the well-managed advance on Komati Poort. The occupation of that place and the complete disorganisation of the enemy as shown by the destruction of their guns are most satisfactory."

Sept. 28 A ceremonial parade took place on the 28th in honour of the King of Portugal's birthday. The Portuguese General

1900 and his Staff were present, and the force made an imposing sight. The Artillery present included THE BATTERY and one pom-pom, the 20th and 85th Batteries, four 5in. guns, and two naval 12prs.

Sept. 29 On the 29th the Right Section and one pom-pom started for the Sabi River with 200 Mounted Infantry. All the fittest of THE BATTERY'S horses were sent, but beyond four days of very hard marching nothing of moment was accomplished. Some lions were, however, met with, and during the night succeeded in carrying off a black and severely injuring a trooper of the escort. A mule was also mauled, and this despite the large fires kept burning.

Troops were now being entrained and sent up to Pretoria as engines were put in order. The last of the Guards left this day.

All men were given quinine daily, and it was proved to be required to combat the malaria. The thermometer was considerably over 100 on 1st of October.

Sept. 30 The Boer laager near the frontier contained an indescribable collection of ammunition and supplies. The senior Officers of THE BATTERY narrowly escaped the accident which injured several of the Infantry working party destroying the explosives.

Oct. 2 At 10 o'clock on the 2nd October orders were unexpectedly received to entrain at once. At 11 the Head Quarters of THE BATTERY, Centre Section, and details were at the Station, but it was 4 p.m. before the two trains started; one contained the Officer Commanding and Section, and the other the Captain

Oct. 3 and transport. Malelane was reached that night, and Kaap Muiden by mid-day on the 3rd. Here it was found that Lieut. R. T. Hill had joined to relieve Lieut. Hine-Haycock, promoted some time since. The latter was appointed to the command of the ammunition column *vice* Major P. E. Gray.

The Left Section entrained and shortly followed us. Leaving Kaap Muiden, the railway passes through a splendid gorge up to Crocodile Poort, and there is no direct road between the two stations. Nell Spruit was reached at midnight.

Oct. 4 Water ran short at the station, and it was not till 5 p.m. on the 4th that the train started, Alkmaar being the limit of the

1900
Oct. 5 journey. On the 5th Waterval Onder was reached early and a rack engine assisted the train up the sharp incline to Boven. There is a difference of 700 feet in the level of the two stations. An early halt was made here and the horses detrained for the night.

Oct. 6 Entraining at 4.30 next morning, good speed was maintained to Machadodorp. Here orders were received to leave all horses and mules, only chargers and one Waler trooper to fill up the truck being retained. Over thirty Walers were handed over, and the Right Section also had some with them.

Olifants River was the limit that day (6th), though reached at 3.30 p.m., but the line near Bronkhorst Spruit had been blown up the previous night.

Oct. 7 The following day one or two shots were fired at the train between Brugspruit and Balmoral, but THE BATTERY safely entered Pretoria at 3 p.m., and, assisted there by teams from the 18th Battery, camped at "Arcadia," one mile to the east of the town.

Oct. 8 On the 8th tents were obtained and utilised for the first time since quitting Bloemfontein six months previously.

Oct. 10 The Right Section rejoined on the 10th, having also given up all their horses at Machadodorp.

Several large packages of comforts and necessaries arrived, sent from England by private donors for the men of THE BATTERY. They had been nine months *en route*, but were most useful and appreciated.

Oct. 14 On the 14th Lieutenant H. W. Hill reported himself, posted *vice* Lieutenant Wood promoted. Lieutenant W. E.

Oct. 20 Clark arrived on the 20th to replace Lieutenant Delves-Broughton, now also a Captain of some standing.

Instead of THE BATTERY proceeding to India, as was anticipated when deprived of all the horses, 170 remounts were drawn from the Depôt on the 17th. They were all horses recently out from England, fat and heavy. 70 mules were also obtained. All the ammunition waggons were horsed for the first time for some months.

Three days later THE BATTERY was ordered out to Commando Nek at an hour's notice, and, to its great regret, finally

1900   parted with Colonel Henry, under whose able command it had been for the greater part of the regular warfare.

Colonel Henry personally expressed his sense of the excellent services rendered him by THE BATTERY, and of the invariable precision of its gunnery.

Brigadier-General Inigo Jones commanded the new column, which, besides THE BATTERY, consisted of

> One 5in. gun,
> The Regiment of Life Guards,
> 2 Battalions of Guards.

It was supposed that the expedition was to capture ex-President Steyn, who was attempting to go south. Moving off at 4 p.m. on Saturday, the force passed through Daspoort, turning westward along the Rustenburg Road. At 9.30, in pitch darkness, bivouac was formed.

◄Oct. 21   The next morning the march was continued to a tributary of the Crocodile River, near Uitvals Nek.

On Tuesday orders were received to return the next day. Not a Boer had been found, and as the remounts had never seen a gun in action, this was perhaps fortunate.

Oct. 24   THE BATTERY reached its camp at "Arcadia" by 9 a.m. on the 24th, and at once prepared for the next day's Proclamation Parade. All chesnut horses were lent to "A" Battery,

Oct. 25   which, with borrowed horses, led the march past in the main square as soon as the annexation proclamation had been read and seven Victoria Crosses presented. THE BATTERY followed, and then some 6,000 other troops.

THE BATTERY was hardly in camp late in the afternoon when orders were received for four guns to take part in another expedition at 3 a.m. that night.

This was under Colonel Cradock, who had 500 Mounted Infantry with him, the object being to clear the valley west of Pretoria. The Right Section remained behind.

In drenching rain the force once more set out along the Rustenburg Road, this time pushing right on to Rietfontein Camp near Commando Nek. The 2nd Battalion Worcester Regiment was there, and gave THE BATTERY a hearty welcome.

1900
Oct. 27 On the 27th, crossing the Crocodile River by the bridge it boasts of, a sharp turn to the south shortly brought the force into the next valley, which was swept in the direction of Pretoria. No Boers were, however, captured, but a great deal of forage was discovered, which was subsequently brought

Oct. 28 in. Bivouac was formed in sight of Pretoria, but it took two and a half hours on the following day to reach the camp.

Oct. 29 A heavy hailstorm on the 29th did much damage.

Prince Christian Victor died the same day, and on the 1st

Nov. 1 November THE BATTERY provided the gun carriage and also led the procession at his funeral.

Nov. 12 On the 12th THE BATTERY joined Brigadier-General E. A. H. Alderson and his Mounted Infantry Corps in the camp outside Daspoort, and here it remained for some little time.

Some twenty Reservists had by now joined THE BATTERY to take the place of casualties ; six of them were transferred to " M " Battery, then about to start for home.

THE BATTERY was still on the Indian Establishment—the families in India, but the men on English pay—an unfortunate financial position, especially as no other Battery was so situated.

Thirty young soldiers also joined at Daspoort, some of them almost recruits. Captain Wood (whose promotion was finally dated the 12th April, *i.e.*, previous to his joining THE BATTERY) was taken into Hospital on leaving and subsequently invalided home.

Captain Delves-Broughton was put in charge of the Depôt

Nov. 19 at Cape Town. From the 19th to the 22nd THE BATTERY was "standing by " ready to move at a moment's notice, but nothing finally came of it.

Nov. 26 On the 26th the Right Section moved out on an expedition to the farms visited on the 27th of October. It returned late on the 28th to find half the camp blown down, and the place

Nov. 28 flooded by a violent storm. The guns had not been in action.

Sergeant Henson, the senior Sergeant, was now ordered home to the Riding Master's course. He had done sterling work throughout the campaign, which had been brought to the notice of the Commander-in-Chief.

1900
Dec. 3　　On account of the attack on a convoy going to Rustenburg the Centre Section and 100 Mounted Infantry went out on the morning of the 3rd of December to Rietfontein (mentioned on the 26th of October). The Left Section and 50 Mounted **Dec] 5 & 6** Infantry also started at 8.30 the same night. The latter returned on the 5th, and the former on the 6th, having only taken the place of other troops pushed on a stage. On the 5th also THE BATTERY took part in Colonel L. A. Chapman's funeral—the Senior Artillery Officer who died in the war.

Dec. 12　　Captain C. A. Sykes arrived on the 12th to relieve Captain Arbuthnot, posted to Brabant's Horse.

Dec. 13　　Major Enthoven, who had been suffering from malaria contracted at Komati Poort, was now ordered by the P.M.O. to Cape Town and back via Durban, and a month's sick leave was granted. On arriving there he was detained for five weeks at Claremont Hospital till a Medical Board ordered him home. He embarked on the 29th January, 1901, and ceased to command THE BATTERY from that date.

When he left Pretoria there appeared no chance of THE BATTERY moving till the new Mounted Infantry in course of training was ready—a matter of some weeks ; but he was hardly clear of the station on the 13th December before news of General Clements' reverse at Neoitgedacht was received, and THE BATTERY went to his assistance.

THE GUERILLA WARFARE.

1900
Dec. 13
CAPTAIN C. A. SYKES assumed command on the morning of the 13th, orders were received at 11 a.m. for THE BATTERY to march at once, and at noon it moved off with the force under the command of Brigadier-General Alderson.

This consisted of the 11th, 12th, 13th, and 14th M.I., and the Canadian Scouts. Although THE BATTERY had been brigaded with almost all the various Australian Corps, this was its first meeting with Canadians.

Dec. 14
Rietfontein was not reached till 11 p.m., but at 1 a.m. the column resumed its march, meeting Clements' force retiring some 7 miles out. The whole returned to Rietfontein, where they remained on the 15th.

Dec. 16
General Clements having assumed command of the combined forces marched at 2.30 p.m. on the 16th to Scheer Nek.

Dec. 17
Dec. 19
Next day Hekpoort was reached. On the 19th the action of Hekpoort (sometimes called Thorndale) was fought.

THE BATTERY (4 guns) worked on the right flank under Colonel Cookson, I.S.C., the Right Section remaining with General Alderson.

At the commencement of the action THE BATTERY galloped across an open plain under a severe cross fire and made for Yeoman Hill, which had just been seized by a few Mounted Infantry. The ground was ploughed, heavy, intersected by a donga and traversed by a barbed wire fence. Bombr. Collar-maker Bailey was hit in the elbow while cutting the wire, Gr. Leonard was shot through the neck, and 2 horses were hit, one of them twice.

The Boers still held the further edge of the flat top of the hill, and THE BATTERY, on coming into action on it, found itself under a cross and frontal fire, the latter being at a range

F

of some 600 yards. Two guns also immediately opened on THE BATTERY with accuracy, but at too great a range for much effect. On the other hand our shrapnel soon subdued the rifle fire from our front, and it was reported that, amongst other casualties, 3 Boers in a kraal had been here killed by one shell. THE BATTERY remained in this position for three hours, and only lost 2 men and 2 horses, Gr. Warrener being struck in the knee and Dr. Hill mortally wounded.

Meanwhile the centre of the Boer position had been raked by a cross fire from THE BATTERY, the Section on the left and the 8th Battery in the centre.

On the arrival of the Inniskillings the hill was cleared, the enemy fled, and a general pursuit commenced. THE BATTERY was in the van advancing at its utmost speed for 7 miles, the Boer left was completely rolled up, and THE BATTERY then came into action to shell the retreating enemy.

The whole action was directed by General French, General Clements having some 6,000 men, while over 20,000 were employed altogether in the endeavour to surround the Boers. Unfortunately, through want of complete co-operation, a pass in the Magaliesberg was not held, and the enemy was enabled to escape without very much loss. Delarey was in command, and had 5 guns and between 3,000 and 4,000 men.

The troops returned in pouring rain to bivouac at Thorndale.

Next day General Alderson congratulated and thanked THE BATTERY on parade, and, on its reaching Pretoria a week later, Colonel Sclater (Commanding R.A. South Africa) also expressed his approval of THE BATTERY'S services on the 19th.

Dec. 20 On the 20th Alderson's force marched right round the Boer position of Boschfontein and found it completely evacuated.

Dec. 21 Marching west next day the enemy was found in a strong position at Naauwpoort. General Alderson at once engaged him, THE BATTERY taking up a good position and shelling the Boers till dark.

It was not till late in the day that the fire was returned by artillery, when a 9 pr. (which must have been present all day) opened on the horses, but did no damage.

1900 Clements' force was rejoined at 9 p.m.

Dec. 22 The combined force marched through Oliphants Nek capturing or burning immense quantities of supplies and forage. THE BATTERY found, and appropriated, 600 lbs. of tobacco.

Dec. 23, 24, 25 The marches on the 3 following days were to Buffel's Hock, Vol Huter's Kop, and Rietfontein, and on the 26th

Dec. 26 Pretoria was reached. Lieut. J. M. Sloan, R.A.M.C., now joined and assumed medical charge of THE BATTERY.

Dec. 31 The last day of the year Lt.-Col. A. V. Jenner took out a column consisting of THE BATTERY, the 13th and 14th M.I., and the Canadian Scouts, to join General Clements to the west.

1901 Marching to Vol Huter's Kop via Rietfontein it was found
Jan. 1 that the Boers had retired from Roodi Kopjes, where it was hoped to engage them.

Jan. 2 & 3 Remaining at Vol Huter's on the 2nd and 3rd, Rietfontein
Jan. 4 & 5 was regained on the 4th, and Pretoria on the 5th.

These long marches, coming so soon after recent exertions, told heavily on the horses.

Jan. 10 THE BATTERY was once more in motion on the 10th, again under General Alderson. Moving to the S.W. some 10 miles along the Potchefstroom road 40 or 50 Boers were found. They fled, and THE BATTERY only fired 18 rounds, the force returning to Pretoria the same night.

Jan. 12 Two days later the same force marched through Quagga Poort to Schwatz Kop, and encamped 20 miles from Pretoria. The enemy's guns were heard, but could not be located before dark.

Jan. 13 On the 13th hands were joined with General Plumer, but the enemy disappeared to the east. At 5 p.m. the return march to Pretoria was commenced and camp reached at 11 p.m., over 33 miles being covered during the day.

Jan. 17 Alderson's command, including THE BATTERY, moved to Eerste Fabrieken on the 17th, leaving the Right Section in Pretoria. It rejoined on the 25th, and the same day the force was augmented by the Right Section of " G " Battery, under Lieut. E. F. Calthrop, and the 2nd Battalion K.O.Y.L.I. This brought the numbers up to over 2,000.

Jan. 26 The march eastward commenced next day, the column forming part of the combined forces that under the supreme command

1901

of General French made the grand sweep to the Ermelo district.

Feb. 6

Mooifontein, within 8 miles of Ermelo, was reached on the 6th February, the various stages being Mooiplaats, Vangatfontein, Bombardy Kop, Deepspruit, Welstand, Schurve Kop, and Nooitgedacht. But little opposition was met with, the fighting being confined to the Scouts, who lost several men and a Colt gun. THE BATTERY fired but a few rounds.

The horses suffered terribly from horse-sickness and bilious fever contracted chiefly at Eerste Fabrieken ; 20 died in a week from the former disease. All foodstuffs, forage and wagons were taken or destroyed, and all the Boer families were brought in. A number of Boers surrendered. Many ponies were captured, and some were used by the detachments, indeed, before the end of the Eastern operations the whole of the detachments were thus mounted.

Feb. 9

On the 9th, the march continued eastward to Goed Hoep, Magdalene Meer, Sand-Kliff, Kliprug, Panbult and Derby.

Some incidents occurred almost daily, such as opposition to the Scouts by a few Boers, burning and destroying supplies, or the capture of cattle and sheep. In one place 5,000 sheep that could not be brought along were killed by the Infantry with the bayonet.

Feb. 17

On the 17th the Left Section proceeded to Wolverkop on detachment, and a small expedition under Major " Gat " Howard, Canadian Scouts, with the Right Section, moved into Swaziland. Major Howard and his orderly were captured by the Boers and deliberately shot. He was known through-out America, had fought and been wounded at the battle of Gettysburg, and had afterwards made his name famous by his bravery and an incident of remarkable heroism with a Gatling gun. Hence the name of " Gat."

The column returned to camp at midnight, having only burnt a few wagons, and much saddened at the loss of their gallant old leader.

The Headquarters remained at Rustplaats Farm, Derby. It rained heavily almost daily, and the only forage available was grass and green mealies. The rainfall had delayed the arrival of a convoy.

1901

During the few days' halt some minor captures were made, amongst which were two wagon-loads of gin !

Feb. 28
March 1
On the 28th, the Left Section having rejoined, the force moved to Piet Retief, and next day to Marienthall. For the next fortnight this camp was a scene of considerable misery. Grazing was the only food for the horses, fresh meat with but little tea or coffee and one biscuit was the ration per man, and it rained more or less continuously.

March 17
March 18
On the 17th the column marched to Vrygunst, and on the 18th to the Pongola River.

March 20
Lieutenant Addison, R.E., constructed a pontoon bridge next day, and on the 20th the river was crossed.

March 22
Some men of THE BATTERY recovered, by diving, six boxes of ammunition lost in the river the previous day, and a convoy of oats at last arrived.

March 23
On the 23rd the force marched to Paul Pietersburg, which was found deserted. Columns under Generals Dartnell, Knox, Pulteney, and Alderson assembled there, the whole under General French.

March 24
March 25
March 26
Next day Alderson's command marched to Welgewonden, and on the 25th the Right Section proceeded with a convoy to Vryheid, returning next day.

Major L. H. Ducrôt arrived and assumed command of THE BATTERY, having been posted some time since, *vice* Major Enthoven.

March 27
On the march to Express on the 27th a few Boers were encountered, two being killed, two wounded, and two captured.

March 28
The march to Manzaan River, though only four miles, occupied the whole of the next day, owing to the difficult ascent of a long, steep hill.

March 29
& 30
March 31
Marching to Pietersrust on the 29th, General Botha's farm at Waterval was reached on the 30th ; and Commandant C. Emmett's on the 31st.

A flying column, including the Left Section and the Section of " G," pushed on again the same evening. Food and forage were carried on the person, guns, and pack mules, and fifteen miles were marched by moonlight.

April 1
On April 1st a nine mile march brought this column into

1901   touch with General Dartnell's; it then returned and later marched yet another eight miles towards Headquarters.

April 2   Next day a part of the column (with the two guns of " G ") returned the remaining seven miles to Emmett's farm, but the

April 3   rest, and the Left Section, only came in on the 3rd, having covered a considerable distance and captured 3,000 cattle.

During this raid each gun and wagon had a team of 10 horses, sometimes augmented to 12, and the three remaining wagons were drawn by oxen.

April 4   On the 4th the whole column marched to Welgewonden,

April 6   and on the 6th through Vryheid to Scheeper's Nek.

April 7   On the 7th to Brak Pan, Zand River, where two squadrons of the 8th Hussars and French's Scouts joined.

April 9   Two days later Spitz Kop was the bivouac.

April 10   A 12-mile march followed; Colonel Jenner continued to advance with his Mounted Infantry, the 8th Hussars and the Section of " G." He engaged the enemy, who retired.

The Right Section, under General Alderson, also shelled the Boers.

April 11   A 10-mile march on the 11th was followed on the 13th by

April 13   one of 15 into Vryheid, five more miles being added in the evening.

April 14 & 16   Moving to the Blood River on the 14th, and Buffalo River on the 15th, Dundee was passed through on the 16th, and the force camped at Glencoe. Here it remained for three days.

Summary   Since December 12th THE BATTERY had marched 800 miles. Horse sickness, disease, and exhaustion had claimed 200 horses, but the men had been remarkably free from sickness. Only one man had died—Dr. Cook, at Rustplaats Farm. In this respect THE BATTERY compared most favourably with any other unit. The equipment continued to give every satisfaction, even the harness being little the worse for the eight weeks' soaking.

General Alderson's column had alone accounted for 11 prisoners, 90 surrendered, besides killed and wounded, which cannot be estimated, 630 horses, 94 mules, 489 wagons, 145 carts, 1,330 trek oxen, 10,000 cattle, 43,000 sheep, 154,560 lbs. mealies, and 78,000 lbs. oat-hay taken, 754,200 lbs. mealies

1901 and 750,000 lbs. oat-hay destroyed, 67 rifles, and 15,850 rounds of ammunition, while 700 women and children had been brought in.

Fifty-five remounts joined on the 19th, two of which died the next day, and in the course of three days 18 more were sick, veldt sores being one of the causes, a disease then very prevalent.

April 20 to 22 From the 20th to 22nd the marches were:

15 miles to Dannhauser,
13 miles to Ingegan,
7 miles to Newcastle

THE BATTERY expected to return to Pretoria, but was now ordered to take part in the operations round Ermelo under **April 27 12 m.*** Brig.-General Bullock, and five days later it marched to Ingogo, the column consisting of

THE BATTERY.
Four guns 74th Battery (Major McLeod, R.F.A.).
Two guns Commander-in-Chief's Body Guard.
One pom-pom Commander-in-Chief's Body-Guard.
One pom-pom (Captain G. Baillie, R.F.A.)
5th Corps Mounted Infantry (Col. Jenner's).
Gough's Mounted Infantry.
Johannesburg Mounted Infantry.

April 28 17 m. Passing over Laings Nek, under the lea of Majuba and Pogwan, Volkrust was reached on the 28th.

April 29 3 m. The column marched out a few miles and formed laager, a formation General Bullock always used when practicable.

Colonel Stewart, Johannesburg Mounted Infantry, was detached with a force for four days to make a parallel movement. In a rear-guard action he had an officer and six men wounded, but next day the rear-guard concealed themselves in a donga and captured General Schwatz and 10 Boers.

Ap. 30, 8 m. On the 30th Zand Spruit was the camp, and next day **May 1, 12 m.** Joubert's farm.

May 2 20 m. On the 2nd of May a part of the force with four guns of THE BATTERY started at 3 a.m., and made a wide detour on

---

* Miles marched.

1901

May 3
7 m.

May 4
7 m.

May 5
8 m.

May 6
8 m.

May 7
14 m.

May 8

May 9

May 10 & 11

May 12
10 m.

May 13
10 m.

the left flank, marching 20 miles in reaching Amersfort. But few Boers were seen.

Next day four guns again made a wide flank movement, and were engaged with the enemy's snipers all day, firing 50 rounds.

On the 4th a 14pr. Creusot fired a dozen rounds at long range, then hastily withdrew; but a pom-pom and snipers continued to annoy the column throughout the day till the Vaal River was reached.

The Centre Section left the same night with a force under Major Gough, and visited Blauw Kop next morning.

The Left Section also made an early start, proceeding with the General at 3 a.m., and assisting in surprising a Boer picquet, 17 men being accounted for.

The remaining Section accompanied the main column, and was also engaged with snipers during the march.

All three columns met in the morning at Vleiplaats.

Next day two guns were with the advanced guard, and 250 Boers were engaged. A severe shelling put them to flight, and 100 horses were captured. During the afternoon a Creusot suddenly opened on the Section at 7,000 yards. Moving under cover, these guns got within 3,800 yards of the Creusot, and shelled it as it retired.

On the 7th the Mounted Infantry and four guns of THE BATTERY made a wide movement of fourteen miles, and again came under fire of the Creusot at 8,000 yards. A pom-pom also was active, and sniping continued all day.

The force did not move on the 8th, but just before dusk the Creusot shelled the camp. Several shells fell in THE BATTERY'S lines, one actually striking No. 6 gun and putting it out of action for two days. No one was injured.

Marching at 2 a.m. next day, 600 Boers were found in a strong position near Ermelo. The guns soon turned them out of it, and the column occupied the kopje, though a pom-pom opened on THE BATTERY as it reached the summit.

The following days were occupied in seizing Ermelo and bringing in families, forage, and so on.

A slight rear guard action took place during the march on the 12th, and a similar affair on the 13th on the way to Kaffir Spruit.

1901
May 14
13 m.

During the march to Tweefontein on the 14th THE BATTERY shelled the enemy's snipers at intervals during the day.

May 15
30 m.

On the 15th General Bullock started with his mounted troops at 4 a.m. to try and surprise the Creusot, but only two wagons were found. These were captured after a chase, and found only to contain families. THE BATTERY was called on to cover 10 miles at great speed, which it did without a check.

During the return the usual rear-guard action took place. Thirty miles were covered.

May 16
7 m.

A convoy arrived on the 16th with Rimington's Column, including the 69th Battery (Lieut.-Colonel Wing's). This force afterwards acted on the left flank.

The "Body Guard" accompanied the empty convoy back to Standerton, the families being sent with it.

A long-range Elswick 12pr. was brought out, which Captain Sykes took command of for a few weeks, till relieved by Lieutenant West, R.F.A., and Lieutenant R. T. Hill also returned after a month's sick leave. In the evening the force moved seven miles.

May 17
15 m.

No Boers could be found next day, and camp was formed five miles from Ermelo.

May 18
20 m.

The force, on the 18th, took up a line from Lake Chrissie to Ermelo, THE BATTERY being near the lake with Gough's Mounted Infantry and the Johannesburg Mounted Rifles, the 74th Battery and Jenner's Mounted Infantry near Mooifontein, and the Elswick gun, Dorsets and transport at Tafel Kop overlooking Ermelo.

This line was held for some days, but there was little doubt that the enemy had broken through,

During the previous month THE BATTERY had again been decimated by horse sickness. For the last week quarter rations only had been issued, and as the cold had been intense the hardship was accentuated.

The enemy had been repeatedly frustrated by General Bullock's sudden changes of plans and his night marches.

THE BATTERY had on many occasions been called on to make very rapid movements, and the horses responded very gallantly.

1901

The enemy were about 1,000 strong with one gun and one pom-pom under the command of General Botha, with whom were Chris. Botha, Smutz, Reitz and Schalk Burger.

May 20 to 24.

The main column did not move till the 25th, but between the 20th and the 24th the mounted troops reconnoitred.

Four guns of THE BATTERY proceeded with Colonel Stewart and his Corps in a northerly direction, and a part of this force with the Right Section penetrated to within a few miles of Carolina.

On leaving Ermelo, Civil Surgeon F. Cornwall took over medical charge of THE BATTERY from Lieut. Sloan.

May 25 40 m.

General Bullock started at 10.30 p.m. on the 25th, taking THE BATTERY and his mounted troops to try and surprise Botha. He moved to the east, but after a 40 mile march

May 26

rejoined the column (14 miles from Ermelo) without having met with any success.

May 27 to 30

Passing through Tobias Smutz's farm Viljeon's Drift was reached, in four marches, by the 30th. There was a great deal of sniping, but little or no opposition.

The grass was now dry and inflammable, and the General ordered the grass on the camping ground to be burnt daily. This caused great discomfort to both troops and animals, and everything became black and filthy.

June 1 20 m.

On June 1st the General Officer Commanding took out a flying column, the remainder moving 10 miles to the Vaal River.

The Centre and Left Sections accompanied the former force, and 120 Boers were engaged. Their losses were unknown, with the unfortunate exception of a woman and two children killed by the pom-pom. They were found in an outhouse which the Boers had been holding.

The men hit were, as was so frequently the case, already dexterously removed.

June 2 10 m.

Next day the whole force marched to Kaffir Spruit, perhaps as trying a day for troops as any in the campaign. A bitterly cold wind was blowing in their teeth, bringing with it the dust from the burnt grass.

June 3 12 m.

The 3rd was a glorious day after a hard frost, and the march was to Tweefontein.

1901

June 4
10 m.

June 5
13 m.

After a slight rear-guard action on the 4th, Grobelaar's farm was reached on the 5th. Though within three miles of Standerton, the General Officer Commanding there would not allow the column to come any nearer to it or the Vaal, and horses had to water in muddy holes.

June 5
to 9

The force halted for five days, during which the Volunteers with the Elswick gun entrained for England, and Lieutenant R. T. Hill left for Pretoria on promotion, Lieutenant M. Hartland-Mahon joining in his place.

June 10
5 m.

June 11
9 m.

Part of the force, including THE BATTERY, moved to Leeuwspruit on the 10th, and was joined by the rest next day (less the 12th M.I.), when the whole marched to Nickersvlei.

June 12
10 m.

On the 12th a few rounds were fired by THE BATTERY at snipers, and the camp was pitched at Morganzon.

June 13
15 m.

The General Officer Commanding next day took a flying column, including THE BATTERY, to Vereeniging, Kaffir Spruit. Only a few shots were exchanged.

June 14
20 m.

The main body on the 14th marched to Dothplaats, and the column rejoined it after a 20 mile march. It may be mentioned that THE BATTERY invariably took part in the night marches, while the 74th Battery and the Elswick gun usually marched with the main body or transport.

June 15
8 m.

A few shots were fired on the 15th on the way to Beginderlijn Bridge, Vaal River.

June 16

No move was made next day till 9.45 p.m., when two flying columns marched, one with the Left Section under Major Gough, the other with the rest of THE BATTERY under the General Officer Commanding.

June 17
28 m.

Taking circuitous routes, they marched through the night. It was freezing hard. But 28 miles were covered by each force ere they concentrated at Rolfontein, where the rest of command had moved in the morning.

June 18
10 m.

A few Boers were seen on the way to Driefontein on the 18th.

June 19
8 m.

A short march to Langkloof was made next day.

June 20

A convoy went into Wakkerstroom, seven miles off, on the 20th and replenished, while |THE BATTERY played a cricket-match against Gough's Mounted Infantry, and won by 28 runs.

1901
June 22, 5 m.
June 23, 6 m.
June 24
4 m.

Mooipoort was the extent of the march on the 22nd, and Spitzkop on the 23rd, but there were wide flanking parties.

There was more sniping than usual on the 24th, and THE BATTERY fired 64 rounds on the way to Rudipoort. One gunner was hit but not hurt.

June 25
18 m.

Next day the Right and Left Sections took part in a raid into the low veldt under the General Officer Commanding.

After a long descent off the mountains many cattle and sheep were captured, and oxen sent to await the return of the guns dragged them up the steep ascent in the evening.

June 26
6 m.

There was heavy sniping on the 26th while the J.M.R. stormed a kopje overlooking Beelzebub Farm, our destination. The Elswick gun at 7,500 yards made good practice.

A helio message from Lord Kitchener here told the General Officer Commanding he was out of his district, and preparations were made to return to Ermelo.

June 27
10 m.
June 28, 5 m.

Marching to Mooipoort on the 27th, only five miles were traversed next day, three bad drifts delaying the transport. Snipers succeeded in wounding Dr. Jones in the knee.

June 29
10 m.
June 30
33 m.

On the 29th Schalp Spruit, near Amersfort, was reached.

Next day the General Officer Commanding, acting on information from Standerton that a meeting of Boer Generals was being held at Blauw Kop, proceeded there with a flying column and four guns of THE BATTERY (Right and Centre Sections).

The Kop was taken, but only a few snipers found.

The column rejoined the transport at Strynek where it had moved to. A long day, and THE BATTERY lost three horses.

July 1
10 m.

The rear-guard was engaged on the way to Wolverspruit on July 1st.

July 2
7 m.

The railway at Platrand was the destination next day, and two days' forage was obtained there.

July 3
12 m.

At Katboschspruit the water was particularly bad.

July 4
7 m.

A short march was made on the 4th to the Vaal at Standerton, and some remounts and stores were obtained.

Here the Centre Section, under Lieut. Clark, left THE BATTERY and went to Rimington's column, not rejoining till May, 1902. A section of the 74th Battery, the Body Guard, and the J.M.R. (less one company) also quitted the command.

1901
July 7
10 m.
Three days later the column moved to near Joubert's Kop, and almost captured a Commandant, the Mounted Infantry getting within thirty yards of him.

July 8
20 m.
The transport took all next day to cross Roberts' Drift. The General Officer Commanding took out a flying column at 2 p.m., and engaged about 50 Boers, some of whom appeared to fall to THE BATTERY's shell. Camp was regained at seven.

July 9
10 m.
A 10 mile march to Inloop followed. The country here was well stocked, mealies, oat-hay, and manna grass, besides pigs and poultry, being found in all the farms. This must all have been imported during the previous two months, whilst no columns were traversing the district.

The country, unlike the Transvaal, abounded in ground game. A few bustard of both kinds and numbers of hares, plover, dikkop, and partridge afforded excellent sport for gun and greyhound.

July 10,
11, & 12
8 m.
Marching to Sidbury on the 10th, and Schiedam on the 11th, a small flying column went out on the 12th. The General Officer Commanding took the Right Section with him, and engaged some 50 Boers who, as usual, fled.

July 13
12 m.
Next day the command marched to Villiersdorp. The farms thereabouts were still well stocked.

July 14
9 m.
Marching at 1 p.m. on the 14th the Vaal was reached, where there was much delay owing to the large number of captured sheep objecting to cross. Nearly 2,000 were finally abandoned. Late in the day Major Gough took the Left Section and the J.M.R. in pursuit of 100 Boers. One Mounted Infantry-man was killed. The camp was at Rietfontein.

July 15
25 m.
Next day another mounted force went out, the General Officer Commanding taking the Right and Left Sections with him. 300 Boers were pursued and shelled at intervals. A mill was broken up and camp regained after dark.

July 16
10 m.
On the 16th the order was reversed, the main body moving off first. The flying column was concealed, and on the Boers arriving, as usual, to search the camp for ammunition, THE BATTERY opened fire. The enemy remained so boldly that at first they deceived the General Officer Commanding; but eventually they received a hot fire.

July 17
10 m.
Another 10 miles next day, past a large German mission

1901

July 18
9 m.

station, brought the force to Zuikerbosch Spruit, near the Nigel gold mine. Heidelburg was reached on the 18th and a clear week's rest was enjoyed, the longest THE BATTERY had been allowed since December 12th.

The remaining company of J.M.R. left to rejoin their Corps.

Brig.-General J. Spens relieved Brig.-General Bullock and assumed command on the 22nd.

July 27, 12 m.
July 28, 10 m.

On the 27th and 28th the column moved to Malan's Kraal and an unnamed " pan."

July 29, 8 m.

After crossing the Vaal next day at Grobelaar's drift, and leaving the transport at Bracklaagte, some 60 Boers were dispersed and cattle and sheep captured. Colonel Jenner's flying column with the Right Section left for Villiersdorp, which he reduced to ashes, and rejoined after covering 65 miles in two days.

July 30 &
31, 65 m.

The main body meanwhile moved five miles to a farm near Van Reenans Kop, meeting a few snipers. Gough's Mounted Infantry and the Left Section left at 3 a.m., recrossed the Vaal, and reconnoitred down the right bank, recrossing the Vaal and Wilge rivers—a 60 mile march.

Colonel Dunlop, C.R.A., also took out a small column and made a successful reconnaissance.

Aug. 1
8 m.

The whole force marched to the Klip river on the 1st, crossing the Wilge.

Aug. 2

Two flying columns again went out during the night of the 2nd.

Colonel Jenner's left at 11 p.m. Captain Sykes accompanied it as C.R.A., the guns being two of the 74th Battery, under Lieut. Norman Ramsay. A Boer laager was found by Major Pratt, 13th Mounted Infantry, at 3 a.m., sleeping and outspanned, and the following captures were made :—35 Boers, 14 Cape carts, 14 wagons, various families, 20 Kaffirs, 60 rifles, and 1,800 cattle. Halting at daybreak, Frankfort was burnt to the ground.

Meanwhile Gough's column with the Right Section started at 3 a.m., and at 6 a.m. found a commando in full flight. It was pursued and overtaken, 15 wagons being burnt, 18 Cape carts and 3,000 cattle being brought in.

1901

Both these captures were effected without any loss.

Colonel Jenner rejoined on the 3rd, and Major Gough on the 5th.

ug. 3, 18 m.
ug. 4, 8 m.

The main body reached Lubby on the 3rd, and Gravenhage on the Klip next day. Only water in puddles was obtainable.

Aug. 5
12 m.

On the 5th the whole column reached Heilbron.

At midnight Colonel Jenner and Major Gough both left with flying columns, the Right Section being with the former

Aug. 6 & 7
23 m.

and the Left Section with the latter. Jenner's force took five Boers, 21 Cape carts, one wagon, 700 cattle and other stores, and with Gough's column rejoined on the 7th.

In the meantime Head Quarters moved to Leeuwfontein on the 6th and Paardekraal on the 7th.

Aug. 10
35 m.

It was not till the 10th that a move was made, and then one of but five miles, all within the limits of the farm.

Jenner's column, with the Right Section, was again out early, captured five prisoners, and returned after covering 35 miles ; Gough's and the Left Section secured two Boers and traversed 20 miles.

ug. 11, 12 m.
ug. 12, 10 m.

Marching to Doornkop on the 11th, Kroonstad was reached on the 12th. Here advantage was taken of the new system of the R.A. Remount Department, and debilitated and lame horses exchanged for fresh ones.

Aug. 16
35 m.

Four days later the column marched 10 miles towards Lindley. At 4.30 p.m. Colonel Jenner set out from camp with a large force including THE BATTERY, and surrounded Lindley by daybreak. This was done without opposition, and the town was burnt to the ground.

Thirty-five miles were covered in 23 hours.

ug. 17, 15 m.
ug. 18, 15 m.

On the 17th the column marched to Doornkop, and on the 18th to Paardeplaats, where Jenner's force rejoined. At 7.30 p.m. some Boers drove in a picquet and sniped the camp. Fires were immediately extinguished and the enemy easily driven off.

Aug. 19

No move was made on the 19th, but THE BATTERY prepared to fire at night with picket posts and lanterns.

Aug. 20
24 m.

Next day the column marched to Lovedale. Gough's Mounted Infantry and the Right Section on the right flank engaged about 100 Boers near the scene of the Body Guard's Lindley reverse, traces of which were seen in profusion.

After marching seven miles on the morning of the 21st, a flying column started at 11 p.m., rejoining 10 miles further on next day. Jenner's force, with the Right Section, and Gough's, with the Left, had surrounded a supposed laager, but it had gone five days earlier. However, 150 Boers happening to be there, were almost cut off, and only escaped by hard galloping through a gap, three being captured. Sixteen saddles were picked up, and from the heavy cross fire to which the Boers were subjected—once at 200 yards— several must have been hit.

A short march followed to Valsch River, THE BATTERY (still only two sections strong) moving with the flanking parties and covering 10 miles.

In pouring rain 10 more miles, to Sidbury, were traversed on the 24th.

A similar experience in thunder storms marked the march to Spitzkop.

On the 26th the two sections were again on the flanks and the usual sniping was experienced.

Two days later the column moved on to Meerkatfontein. Captain Dick, on the extreme left, pushed on so rapidly that he became detached from Gough's three remaining companies. He ran into a Boer force guarding a convoy, but held his own gallantly until the other companies and the Right Section galloped up and dispersed the Boers. The Mounted Infantry lost one officer and three men killed and 12 men wounded, but from various reports the enemy appear to have suffered at least as much.

Next day the column marched to within four miles of Kroonstad.

After two days spent in replenishing stores the force moved to Klumpdoorn.

At 5.30 a.m. on the 2nd Colonel Jenner took out THE BATTERY with a flying column, and sighting a convoy galloped in pursuit. He captured 15 wagons, 7 Cape carts, and 6 prisoners, driving the remainder into Rimington's column, where 51 wagons, 22 Cape carts, 18 prisoners, and many cattle were secured.

The operation was planned by General Spens. Colonel

1901 Jenner returned to camp, now at Grasfontein, having covered 40 miles. A search light in camp greatly assisted the column to find its way in the darkness.

Sept. 3 The force marched to Jagersfontein on the 3rd, and
10 m. Welgegund on the 4th. Here two wagons and three Cape
Sept. 4 carts were found, and Major Gough at once went in pursuit of
8 m. the commando that had abandoned them. He took a pom-pom
Sept. 5 but no guns with him, and returned next day with 18 wagons,
60 m. 29 Cape carts, and 4 prisoners.

The same day Colonel Jenner started with the Right Section at 12.30 a.m. into the Doornberg, returning on the 7th. Only two Boers were captured.

Excluding the 60 miles just finished, THE BATTERY had completed the following distances :—

December 13th, 1900, to April 28th, 1901,
with General Alderson ... ... 800 miles.
April 29th to July 18th, 1901, with General
Bullock... ... ... 780 „
July 27th to September 4th, 1901, with
General Spens ... ... 480 „

2,060 „

Average march with General Bullock ... 14 miles.
„ „ „ Spens ... 16 „

Sept. 6 Head Quarters halted. Three hundred fish were killed in the spruit with dynamite, but proved very bony and muddy.

Sept. 8 & 9 On the 8th 12 miles and on the 9th 11 were marched,
23 m. while next day Kroonstad was reached. Here, to the regret
Sept. 10 of everyone, Gough's Mounted Infantry entrained for Natal,
12 m. to oppose Louis Botha.

While the column was halted for a week the Dorset Regiment left for Brandfort. Cricket and dancing were indulged in, the former played to the sound of desultory firing from the forts.

Sept. 18 The column was once more in motion on the 18th, moving
10 m. towards Bothaville, and camping 10 miles out. Orders were

G

1901   here received to return at once on account of the disaster to
          Gough's force.

Sept. 19   The whole column was back in Kroonstad by midday on
10 m.   the 19th, and entrained by successive regiments. Dundee was
Sept. 22 reached and the detraining completed by 10 a.m. on the 22nd.
          The traffic was much congested by several columns being
          pushed towards the same station.

Sept. 23   At 11 p.m. the column commenced a night march, its
16 m.   strength having been augmented by four companies of the
          Cameron Highlanders.

              Leith was reached after a very arduous march, some of the
Sept. 24 wagons only making camp on the 24th. The column reached
15 m.   within two miles of Rorke's Drift. The crossing of the drift
Sept. 25 next day occupied six hours, as the Buffalo River was much
          swollen by rains.

Sept. 26   On the 26th nine miles to Vant's Drift were marched, and
9 m.   hands joined with Colonel Allenby's column. These two
          columns, together with Colonel Pulteney's, now came under the
          command of Major-General Bruce Hamilton. All mule wagons
Sept. 27 were sent into Dundee, returning at 2 p.m. next day reloaded.

              News having been received of the siege of Itala, a mobile
Sept. 28 force selected from the three columns started at midday for
45 m.   its relief. The Left Section accompanied it, and 45 miles
          were covered in 23 hours. The garrison was found to have
          evacuated the place, having driven off the enemy with loss,
          though five times outnumbered. The force halted near
          Babananga and awaited its transport. This started at
          4 a.m. on the 28th, escorted by the Right Section, 1 Section
          R.F.A., and a mixed force of Scots Greys, Carabiniers,
          Victorians, Gough's Mounted Infantry, Cameron High-
Sept. 29 & landers, and Cameronians. It was believed that the Boers
30   would attack; hence this force of 1,400 men, which Colonel
          Dunlop commanded.

              Passing over the historical battlefield of Isandhlwana,
          Babananga was safely reached on 30th.

Oct. 3,   No move was made till the 3rd, when camp was shifted
5 m.   five miles to the high ground overlooking Itala.

Oct. 4,   Next day three mobile columns started, Spens' column
12 m.   marching to Bondlyst.

1901   The country through which the Zulu border runs is very
beautiful and luxuriant. The farms had been unmolested and
were very pretty, several were passed containing severely
wounded Boers. The water was good, but the going very
mountainous.

Oct. 5, 11 m.   On the 5th Klipstapel on the White Umvelosi was reached,
Oct. 6, 12 m. Uitzicht next day and Bevenson on the 7th.
Oct. 7, 15 m.   There was a long climb up to Thabankulu two days later,
Oct. 9, 8 m. and snipers were about.
Oct. 10, 10 m.   Turning back through Bevenson on the 10th, camp was
pitched at Ekuhlengeni on the Black Umvelosi.

The reason of this counter-march was that Boers, families
and wagons were reported to be in the Ngomi Forest to the
East. With Allenby's column, they were to be pursued
instead of Botha's force retreating to the North.

Oct. 11   Next day the heights to the East and North were to be
taken, but camp was to be regained the same evening. The
troops accordingly were lightly equipped.

In the valley some of the worst dongas yet seen were
encountered, and the ascent finally required 18 horses in each
gun supplemented by 40 men on drag ropes. Some guns had
10 horses and 6 mules, others had part of the ammunition
carried up by hand.

On reaching the summit Boers driving cattle were seen; the
force divided into two parts and gave chase, but heavy rain
stopped the advance and the troops bivouacked in the open
without food or blankets.

Oct. 12, 18 m.   Next morning one portion kept to the left over the Inyati
Nek, where a sniper wounded Dr. Back in the foot. The
remainder kept to the high ground and rejoined at Allenby's
Camp without meeting with anything. Man and horse had
been practically without food for 36 hours. Allenby's force
showed generous hospitality, and THE BATTERY were indebted
to " O " for sharing their scanty rations.

Oct. 13   A halt was made at Welkern, and on the 13th the baggage
rejoined after an absence of 10 days. While here patrols and
outposts were persistently sniped, and Major Ducrôt left for
Vryheid with Allenby's column to obtain supplies.

Oct. 16, 15 m.   On the 16th the force marched to Rietvlei, from whence

1901   Major Gough surprised a laager. A lady sniper in light attire was all but captured.

These trying days, during which horses frequently had barely 3lbs. of forage, depleted the three columns of 800 horses, and it became necessary to obtain fresh ones.

Oct. 17   Accordingly next day a force consisting of representatives of each unit, under the command of Captain Sykes, left for Dundee to draw remounts.

Oct. 19, 13 m.   The column marched to Vryheid on the 19th, and on the

Oct. 20, 15 m.   20th a mobile force about 800 strong, including the Right Section, left under General Spens for Paul Pietersburg.

Oct. 21, 12 m.   The column entered Pietersburg on the 21st and exchanged

Oct. 22, 12 m.   convoys, returning towards Vryheid next day. Eight Boers were captured.

Oct. 23, 15 m.   Marching to Vryheid on the 23rd two more were taken.

The force that had remained at this place, including the Left Section, marched the same day to the Blood River.

Oct. 24, 13 m.   The mobile column followed next day and caught it up on the

Oct. 25, 13 m.   25th at De Jager's Drift.

Oct. 26 to 28, 41 m.   The next three marches were to Dannhauser, Allcock's Spruit, and Newcastle, and during the following days the column was thoroughly renovated. The 1st Black Watch now joined in place of the Cameron Highlanders.

Nov. 2, 14 m.   Restarting on the 2nd, the force crossed Botha's Post and camped below Botha's Pass, the scene of the Ingogo action in 1881.

Nov. 3, 9 m.   Crossing the Pass next day there was a good deal of sniping. Gough's Mounted Infantry and the Right Section left at 11 p.m. to surprise the Boers. Three of the Mounted Infantry, who had stopped to girth up, were captured.

Nov. 4, 13 m.   Marching to Giddyshoek on the 4th a halt was made on the following day on account of heavy rain.

Nov. 6, 15 m.   Brakpan was the camping ground on the 6th. Colonel Jenner worked to the left, and, taking the Left Section with him, was away till the 8th.

The main column was heavily sniped on a pass; the Boers were exceedingly bold and continually charged isolated men, taking several prisoners. The Provost Marshal had a narrow escape when visiting a farm, and was only saved by the fire of

1901 the guns. One of his men was wounded and two were captured.

Some hundred Boers were engaged during the day, and the Right Section fired 100 rounds.

Nov. 7, 10 m. Marching back to Roedehoek on the 7th the Boers were again active, and captured a man or two.

Nov. 8, 10 m. The column marched next day towards the Wilge River, near Vrede. A big surrounding movement was being organised from Pretoria, the columns employed being Dartnell's, Lowe's, Broadwood's, Brigg's, Byng's, Damant's, Rimington's, De Lisle's, and Spens'.

Nov. 9, 12 m. Gough's Mounted Infantry and the Right Section left on the 9th for two days. The column camped seven miles from the Wilge, Major Pratt, assisted by the Left Section, taking

Nov. 10 Bothasburg on the right. In the evening a force of Infantry, Mounted Infantry, and a pom-pom marched to the Wilge to block the drift.

Nov. 11, 8 m. The column crossed the river next day and encamped on the left bank. The rearguard was sniped, but a gun of the 74th Battery killed a Boer, and a prisoner was drowned whilst crossing the drift.

Nov. 12, 10 m. The concentration of the columns was a magnificent sight, all the helios being visible simultaneously on the 12th ; but a gap of 12 miles permitted De Wet, 500 Boers, and 12,000 cattle to escape. About 50 of the enemy were accounted for. Sergt. Calvert almost stepped on two Boers as they lay concealed under the ruins of a Kaffir hut. They were secured.

Nov. 13, 15 m. Recrossing the river on the 13th the force encamped five miles N.E.

Nov. 14, 13 m. Marching towards Standerton, 28 miles were made good
Nov. 15, 15 m. during the next two days. Major Gough on the right saw a good number of Boers, the Right Section being with him. Colonel Jenner left on the 15th for a raid *via* Joubert's Kop and Robert's drift.

Nov. 16, 14 m. The column marched to the Klip River and crossed by De Lange's drift with about 2,000 captured cattle and 15,000 sheep.

Nov. 17, 14 m. Standerton was reached on the 17th, and Colonel Jenner rejoined. The force remained here for 10 days, but the

1901   majority of the mounted troops were employed in escorting convoys to blockhouses.

Nov. 28, 10 m.   The column marched along the line of blockhouses—these miniature fortresses destined to play so important a part in ending the war.

Nov. 29, 13 m.   Moving next day to Protest, camp was pitched alongside Colonel Rawlinson's column.

Nov. 30, 15 m.   The camps following were at Plakfontein, Drinkwater, on **Dec. 1, 13 m.** the Ermelo road, and Ermelo. At 7.30 p.m. on the 3rd the **Dec. 2, 13 m.** Right Section accompanied Colonel Jenner's force and another, **Dec. 3, 14 m.** all under General Bruce Hamilton, on a 22 mile march. At **Dec. 4, 27 m.** dawn two laagers were surprised at Oshoek, 93 prisoners taken, one Boer killed, and 28 wagons, 33 Cape carts, 1,500 cattle, and quantities of sheep and stores captured. The forces marched five miles towards the blockhouse line and encamped.

Dec. 5, 14 m.   The remainder of Spens' column, including the Left Section, marched 14 miles to Morganzon, where it found Colonel Jenner. This was the Headquarter camp till the 18th.

Dec. 7, 15 m.   On the 7th a force of 500 Mounted Infantry and the Left Section under Lieut.-Colonel Dunlop, with three days' supplies in Cape carts, left at 2 a.m. for Blauwkop. Some National Scouts, familiarly known as " Tame Boers," were with the troops. THE BATTERY was in action and three Boers were wounded; the Vaal was crossed and Schoolplaats reached.

Dec. 8, 6 m.   Next day the force returned to the Vaal, but before mid-**Dec. 9, 30 m.** night left again for Kaffir Spruit. At Hollandia two Boers were captured and 1,200 cattle, 600 of which were trek oxen. Morganzon was regained the same night.

Dec. 11, 16 m.   On the 11th the Right Section and 500 men, under Major Pratt, 13th M.I., left to co-operate with Plumer. They marched to Hollandia, via Majuba, and camped.

Dec. 12, 18 m.   Next day, after crossing the Vaal, 1,000 cattle were captured and Kat Spruit was reached.

Dec. 13, 12 m.   On the 13th the force returned to Morganzon; 15 Boers with cattle surrendering at Blauwkop.

Dec. 14   General Spens left at 11 p.m. on the 14th, with a force including the Left Section, to co-operate with Colonel G. G. Simpson, R.A. The former surprised a farm, captured 15

1901	prisoners, and drove three more into Colonel Simpson's column. To the regret of all ranks Major Gough was severely wounded in the arm.
Dec. 19, 11 m.	The force remained halted till the 19th, when the 14th Mounted Infantry were surprised and overwhelmed by a superior force of Boers. The column therefore crossed the
Dec. 20	Kaffir Spruit and halted 11 miles on. Next day the dead were buried. Colonel Jenner marched out at 11 that night, but failed to surprise the enemy.
Dec. 21, 9 m.	On the 21st the column moved to Beginderlijn Bridge, where it remained two clear days.
Dec. 24 & 25 24 m.	Vaalbank Spruit drift and Amersfort were the camps on the 24th and 25th.
Dec. 26	Next day Christmas festivities were indulged in; sports were held, in which THE BATTERY was very successful, winning, amongst other things, the tug of war and mounted wrestling. A camp fire and "sing-song" under arms followed.
Dec. 27, 23 m.	War was recommenced next day, when the Left Section and 400 Mounted Infantry moved out under the command of Major Ducrôt to surprise Zooalshelykt farm. This he successfully achieved, capturing seven and wounding three Boers. He also took many rifles, bandoliers, greatcoats, and so on, lost on the 19th.
Dec. 28, 10 m.	The column (which had meanwhile marched to Schoolplaats) was rejoined next day by Ducrôt's force.
Dec. 29	At 11.15 p.m. on the 29th General Spens left with the Right Section and 450 Mounted Infantry to co-operate with General Plumer.
Dec. 30, 50 m.	A party of Boers was roused and pursued. They ran into Plumer, were headed back, and, after a fine gallop, 20 burghers and three armed natives were captured. Plumer took three burghers. General Spens rejoined the column at Schoolplaats, having covered 50 miles.
Dec. 31, 9 m.	There was an unopposed march next day to Amersfort.
1902. Jan. 1, 16 m.	On New Year's day the column marched to the Vaal, crossed by the Beginderlijn Bridge, and encamped on the height close to the right bank.
Jan. 2	Next day Colonel Jenner, with 450 men and the Left Section, marched off to co-operate with General Plumer.

1902

Jan. 6, 60 m

The remainder of the column entrenched. Plumer had a severe engagement, but on Jenner appearing the Boers immediately fled. The force covered 60 miles, and rejoined Spens on the 6th.

On this date orders arrived from Lord Kitchener for THE BATTERY to proceed to Elandsfontein, there to be converted into mounted rifles. General Spens addressed the men on parade in most cordial and flattering terms, and further published the following order :—

"BEGINDERLIJN BRIDGE, Jan. 1st, 1902.

"The Brigadier-General Commanding wishes to express his thanks to Major Ducrôt, Officers, N.C.O.'s, and men of 'J' Battery, R.H.A., for the excellent work done by them while serving under his command. 'J' Battery, R.H.A., has distinguished itself by its smartness and efficiency both in the field and in camp, and the accuracy of its shooting has always been of the highest standard. The Brigadier-General much regrets the departure of THE BATTERY from his command and the reason which has made its departure a necessity, but he feels confident that in its new sphere of action 'J' Battery, R.H.A., will fully maintain its already high standard of efficiency."

Jan. 7, 18 m.
THE BATTERY said goodbye with the deepest regret to the General and all friends in his column, and, escorted by the 14th M.I., marched to Morganzon. They were relieved by two guns of the 81st Battery and two of the 8th Battery.

Jan. 8, 17 m.
An easy march along the Standerton-Ermelo blockhouse line brought them to Leeuwspruit.

Jan. 9, 9 m
THE BATTERY reached Standerton and entrained forthwith, taking with it all its guns, harness, and equipment. It kept its heavy horses, and drew rifles and bandoliers. The train left the platform at 2 p.m., and reached Elandsfontein Station at 11.15 p.m.

There being no immediate necessary to detrain, the train was shunted into a siding, and all hands slept till dawn.

Jan. 10
THE BATTERY detrained all horses and rifle equipment about

1902 5 a.m., but left guns, wagons, and harness to be sent on to the Ordnance Department, Pretoria.

THE BATTERY encamped close to the " Rose Deep " gold mine, alongside " P," " T," and " O " Batteries. " G " and " R " were to follow later. Elandsfontein, though a big junction, was an inconvenient place in which to mobilize, as the equipment required had to be drawn from Pretoria, two hours distant by rail.

The whole force, *i.e.*, the Royal Horse Artillery Mounted Rifles, was joined by a pom-pom, commanded by Captain R. T. Hill, R.F.A., who had so recently left THE BATTERY. A Colt gun arrived, was attached to THE BATTERY, and placed in charge of a Corporal. N.C.O.'s, Gunners, and Drivers were instructed in the ·303 rifle.

Jan. 11 On the 11th "O," " P," and " T " Batteries, having been Mounted Rifles for a week, left for Heilbron.

THE BATTERY remained at Elandsfontein until the 18th, during which time it was busily engaged in musketry and scouting. The men showed great keenness in their new *rôle*, and soon picked up the requisite knowledge. In shooting they achieved, after four days' practice, some remarkable results.

Under service conditions, *i.e.*, firing after galloping and walking dismounted, 80 men firing two shots at 800, 600 and 400 yards at a 6-ft. target with a 2-ft. bull, showed 90 hits, including 15 bulls. Colonel Sclater came from Pretoria to inspect THE BATTERY, and expressed great approval.

Jan. 18 THE BATTERY being considered ready to take its place in the field in its new form, entrained once more, and arrived in pouring rain at Heilbron. It was met at the station by Captain E. S. Nairne, just appointed Adjutant of the R.H.A.M.R.

Jan. 19 On detraining next morning two Sections of " R " Battery were found which had also arrived during the night.

Lieut.-Colonel J. Keir arrived the same morning and assumed command of the force.

He appointed Captain Sykes Intelligence Officer and Provost Marshal.

The Right Section left the same day in the new *rôle* of

1902　escort to a convoy; their direction was along the blockhouse line to Frankfort, and on the following day the remaining Section of " J " and the two Sections of " R " followed them.

Just before marching off Bombardier Devan, whilst charging the magazine of his rifle, shot himself through the foot. He was left behind, and soon after invalided home. He was much missed by the Officers, for whom he had catered as Mess Butler, often under the most adverse circumstances, with great zeal.

Jan. 20, 12 m.　The first march was to Brakspruit, the second to Frank-
Jan. 21, 20 m.　fort, and another took the force to Tafel Kop.

Jan. 22, 20 m.　THE BATTERY was still divided numerically into Sections ; but further organization was as follows:—Groups of five were formed consisting of two gunners and two drivers with a N.C.O. in charge, who was always responsible for the movements of his group. These N.C.O.'s were responsible to a sergeant, otherwise THE BATTERY system was unaltered.

Groups were found to be the most convenient units for Cossack posts, guards, and outposts.

Jan. 23　The Right Section having rejoined, " J " and " R " Batteries accompanied Colonel Rawlinson's Column in a night march. The force surprised 21 Boers, all fast asleep in one room, and captured them before they were fairly awake.

Another farm on a flank was surprised, but eight Boers managed to get on to their horses and to head for the open veldt. THE BATTERY ran down these and captured several, driving the remainder into Colonel Scott's (R.H.A.) Mounted
Jan. 24, 42 m.　Infantry. 300 cattle were also taken, and the force, after two hours off-saddle, returned to Tafel Kop, the rear guard being engaged in keeping off Boers throughout the march.

Jan. 26, 12 m.　After a day in camp, the Mounted Rifles marched to Cyprus. Driver Robertson unluckily shot himself through the leg from knee to heel and endured great agony from the wound. He was invalided home.

Jan. 27 & 28,
22 m.　Two marches of 12 and 10 miles brought the two Batteries to Dundas, where "P," " T," and " O " joined them, making a force of about 500 men. The whole left at 9.30 p.m. to try and surprise a Boer laager, about which information had been obtained, in co-operation with Damant. The co-operation was

1902 effected successfully, but owing to the enemy having shifted camp during the previous day it was found by the extreme right flankers. A few rounds gave the alarm and, with incredible swiftness, the Boers saddled up and galloped off, hotly pursued by the R.H.A.M.R. The enemy had a fair start and, on a good position, turned and poured a somewhat destructive fire into " O " Battery, wounding Lieutenant Gill

Jan. 29, 60 m. and several men. The chase was carried on for some miles, but after their long night march the R.H.A. horses could carry their riders no further.

The whole force returned to Frankfort, and the rearguard was vigorously assailed by 200 Boers, part of the 800 which had formed the laager under Commandants Ross and Albertz.

The Battery, acting as rearguard, kept the enemy at bay, with the loss of but one man to themselves. The two Sections held the hills alternately.

The R.H.A.M.R. lost one officer and nine men wounded, and three taken prisoners. They captured five Boers, besides inflicting slight loss. They covered over 60 miles in 27 hours.

One of the Boers was in khaki. He was tried and subsequently shot.

Jan. 30 & 31 The force remained for two days at Frankfort.

Feb. 1, 13 m. On the 1st the force was augmented by Damants' and Wilson's columns, Colonel Keir assuming command of the whole and marching them to Windfield.

Feb. 2, 10 m. Next day the force marched to Van Reenan's Kop, capturing a despatch rider.

Feb. 3, 18 m. The force marched back on the following day to Frankfort,
Feb. 4 and remained there on the 4th.

On this day orders were received to take part in the first of the great drives planned by Lord Kitchener, and the duty of the R.H.A.M.R. was to advance along the blockhouse line, supporting it as the driving columns swept along.

Feb. 5, 10 m. This entailed two marches of 10 miles each and a third of
Feb. 6, 10 m. 20 miles, the bivouacs being Krom Spruit, Brakvlei and
Feb. 7, 20 m. Rietspruit.

At Rietspruit the left of the R.H.A.M.R. was resting on Heilbron, and as the drive was to finish on the following morning it was expected that the Boers would endeavour to

1902 break through. This surmise proved correct, and two very heavy attacks, amongst others, were made upon the Heilbron corner and fiercely repulsed.

Two armoured trains with searchlights assisted. A 4·7 was firing short shrapnel, and rockets were being sent up to indicate points requiring support.

The continuous roll of the machine guns and musketry, the booming of the field guns and the general clamour of the battle made a magnificent effect.

In the morning at the Heilbron Corner 10 dead and 20 wounded Boers were found, besides 50 who had declined to make any further attempt to break through the line, and were sitting under cover waiting to be taken. About 20 heroic Boers appear to have broken through.

Feb. 8, 4 m. The R.H.A.M.R. marched only four miles next morning to Gottenburg, and then all the columns converging, finished the drive on the main line. Although 286 Boers were captured, the result was at the time considered a disappointment. De Wet escaped by his usual expedient of driving cattle in front of him, and slipping through in the confusion.

Feb. 9, 13 m. Next morning Keir's columns marched to Heilbron, and rested there until the 13th.

Feb. 14, 52 m. On the 14th the second drive commenced, and THE BATTERY, by marches of 10, 10, 14, 8, and 10 miles, reached Tafel Kop once more. The bivouacs were at Brakvlei, Kromspruit, Frankfort, Dundas, and Tafel Kop, and at each of these places THE BATTERY, as usual, " dug in," as the men called it. The front defended by a battery sometimes exceeded a mile, and at Tafel Kop it was considerably more, which enabled a few Boers to pass through the line. The number of men to the mile during the night in these drives always varied according to circumstances, and may be said to have ranged from 400 to 100.

Feb. 22, 16 m. On the 22nd Keir's columns left the blockhouse line and formed part of the driving line, bivouacing that night at Bomboes Spruit.

Feb. 23 to 27
51 m. Marches of 8, 10, 18, and 15 miles through Hol Spruit (23rd and 24th), Cornelis Spruit and Kalkoen Krans to Harrismith, finished a most successful drive, which realised 817 prisoners, 200 wagons, 28,000 cattle, and countless sheep.

1902   De Wet again forced his way through by his usual method, in which affair the New Zealanders earned undying fame.

The Spruits were found full of such things as wagons, Cape carts, and furniture, all of which were either burnt or brought along. Though THE BATTERY had little fighting during this second drive, the strain was most trying. Keeping touch throughout a march made the pace slow, and bivouacs were only reached in the evenings, when it was necessary to dig trenches.

As no fires were allowed after dark, this left very little time for cooking dinners. In addition to this, during the last few nights firing at some point in the line had been so incessant, and alarms so frequent, that refreshing sleep was out of the question.

The following telegram received from the Lieut.-General Commanding the forces in the Eastern Transvaal, was published by Brigadier-General Alderson :—

" RUSTPLAATS, Feb. 25, 1901.

" Please let your troops know how highly I appreciate their spirits and bearing under privations they are suffering from bad weather and short food."

The columns rested at Harrismith until March 2nd.

Mar. 2   THE BATTERY left Harrismith on the 2nd, and camped by
Mar: 3   the Wilge River, halting on the 3rd. Major Ducrôt and Captain Sykes were away on sick leave at Durban till 16th.

Mar. 4 to 6   THE BATTERY marched for three days, reaching Tafel Kop on the 6th. Very few Boers were seen, the country traversed having just been swept by a big " drive."

Mar. 7 to 13   From Tafel Kop THE BATTERY marched to Leeuwport, six miles north of Vrede. Turning south, Vrede was passed three miles to the west, and the Gemsbokberg Mountains were worked through and carefully searched. These were very steep, and in many places there was great difficulty in even leading the horses. Only a few sheep were found.

Mar. 14 to 16   Next day the Klip River was crossed, and the column proceeded to Volksrust, reaching it, by Alleman's Nek, in two marches. Here THE BATTERY entrained for the west.

THE BATTERY detrained at 3 a.m. at Klerksdorp, and remained awaiting the massing of all the columns despatched to avenge the capture of Lord Methuen.

While at Klerksdorp, Lieutenant Hartland-Mahon left THE BATTERY sick, and was shortly afterwards invalided home.

Damant's force having been placed at the disposal of Colonel Cookson, I.S.C., Colonel Keir was given the 28th M.I. in its place. The latter's three columns formed part of seven under the command of Major-General Walter Kitchener.

Colonel Rawlinson's columns were also encamped at Klerksdorp, and those of Colonels Kekewich and Rochfort, destined to take part in the combined operations, were lying at some distance north and west respectively. The Boer picquets were seen within 20 miles of the town, and were once or twice encountered by our reconnoitring parties. The main body of the enemy were, however, at least 30 miles off.

At 6.45 p.m. on the 23rd all available mounted men of all these forces, numbering over 10,000, marched silently from their several camps to points indicated to them on the map. Lord Kitchener was himself at Klerksdorp and directed the movements.

The scheme was to pass through the Boers, form up on the western side in a semi-circle, and sweep the enemy into the Klerksdorp-Lichtenburg blockhouse line. Throughout the night a quiet trot was maintained, the Boers possibly imagining the forces had passed them in error. The moon shone out upon the rain-clouds, producing the beautiful effect of the lunar rainbow. At 5 a.m. on the 24th the R.H.A.M.R. formed up in line on a ridge 42 miles from Klerksdorp, the horses seemingly as fresh as ever.

Three halts only, of five minutes' duration each, and a quarter of an hour at dawn were indulged in ; but from dawn
to 7.30 p.m. on the 24th the troops moved incessantly, occasionally at the gallop.

The drive was carried out, but a gap occurred through which a great number of Boers escaped. It must be remembered that in such an operation guiding was everything, and that though there were many excellent guides who knew the

1902	country by day, there were few who could be considered reliable by night.

The drive may be considered a success, as it yielded 178 prisoners, three guns, and two pom-poms.

The R.H.A.M.R. covered no less than 100 miles in the 24 hours, and only lost one horse. Through some mistake on the part of the staff, General Kitchener's force did not find their camp readily, and wandered some miles looking for it. They were overtaken in the evening by a terrific hailstorm, and arrived in camp soaked to the skin.

Ten men of " J " and 20 of " R " became detached during the drive and lost themselves in the mimosa bush which abounds in those parts. After an exciting time they found their way back unharmed to Klerksdorp.

Mar. 25	The troops, after halting next day at Oorbietjesfontein,
Mar. 26, 15 m.	marched back to Klerksdorp, and received the cordial thanks of Lord Kitchener for their exertions.

Lieutenant H. W. Hill of THE BATTERY conveyed a letter to General Delarey, rejoining the force at Oorbietjesfontein,
Mar. 29, 16 m.	where it had again arrived on the 29th.

The columns marched next day to Rietvlei. On the
Mar. 30, 12 m.	following morning at 3 a.m. Cookson's and Keir's columns, numbering about 2,000 men, four guns R.F.A., and three pom-poms, started on a reconnaissance in force. At dawn a convoy, apparently escorted by 500 Boers, was sighted and at once pursued.

The Canadians formed the rearguard to protect the light baggage, which followed on behind.

The R.H.A.M.R. were on the left, and were galloping steadily in three lines. When seven miles had been covered, the force was suddenly brought up dead by a heavy fire from the Boers, who had turned. Damant lost about a dozen men in the first few volleys.

The Boer convoy escaped, but during the delay caused by this fighting the head of the baggage arrived. Suddenly large numbers of the enemy were seen to be galloping round our flanks, and it became imperative to get the wagons in as rapidly as possible. Part of the 28th M.I. were sent out to guard the more exposed flank. In spite of the fact that several

wagons wandered off the direct line, they were brought
into safety. The last few wagons were hardly out of
danger when three of the enemy's guns suddenly boomed
out, and three shells dropped in the midst of the troops;
at the same time a heavy musketry fire broke out from all
directions in a complete circle, and it was obvious that the
British were surrounded by a force far superior to their own.

Without the slightest confusion the troops took up defensive
positions. The Canadians under a terrific fire began making
rough shelter trenches; the R.H.A.M.R. were partly in a
small spruit and partly in the open, lying on the ground.
The position was quite flat, but with a fine open field of fire
in all directions. The Boers occasionally made desperate
charges, hoping to rush the battle, but they were repulsed in
every case. As each charge was made, so was the threatened
side quickly reinforced. One of these attacks was directed
against a farm 1,000 yards away, thought by the enemy to be
unoccupied. But Colonel Keir had placed " T " and " P "
Batteries Mounted Rifles to defend it, and the Boers, when
they had been allowed to get within a few hundred yards, were
met by a withering fire, which compelled them to retire. In
the meantime the three Boer guns, assisted by two pom-poms,
kept up a rapid and continuous fire. Though their ranging was
excellent, their fusing was contemptible, and the British guns
soon ceased duelling with them and addressed themselves to
pouring shrapnel into the Boer riflemen.

The battle raged without abatement for two hours, when,
after one final charge, the Boer leaders could not induce their
men to again face the immovable British force. They vanished
as rapidly as they had appeared, and the scene of terrific din
became suddenly silent.

The Boers afterwards admitted to having over 3,000 men,
but they failed to make any impression upon the courage of
Mar. 31 their foe. The behaviour of the men was magnificent, and at
times they shouted to the enemy to come on. Our losses were
about 200 killed and wounded, and over 500 horses and mules
and that of the enemy, judging from the movements of their
ambulances, cannot have been much less.

Part of the 28th M.I. and some 20 Canadians who were

1902 outside the British circle suffered most, and died heroically. The R.H.A.M.R. lost 24 killed and wounded.

THE BATTERY lost one killed (Gr. Hayes) and two wounded. Though seven gunner officers were wounded only one was dangerously hit.

Fearing a return of the attack, Colonel Cookson sent the wounded to a farm three miles off, under a flag of truce, where they were visited by the Boer Generals, who could not conceal their disappointment at failing to capture the columns. One or two of these said that they would take the British force before dawn. They would have found their enemy deeply entrenched, but with their spirits higher than ever.

The Boers did not make any further attack, and long before the arrival of a relieving column, under General W. Kitchener, they had entirely disappeared from the vicinity. This action was known as the Battle of Boschbult. The R.H.A.M.R. were described in despatches as "behaving with great steadiness, allowing the enemy to get within 200 yards, and then repulsing them with steady rifle fire." They were further specially mentioned in despatches by Lord Kitchener, and congratulations were received a few days after from Lord Roberts and Lord Kitchener in the following words :—

"Extract of a wire received from the Commander-in-Chief, South Africa, by the General Officer Commanding :—

DRIE KUIL, April 2nd, 1902.

Yours just to hand. Regret losses, but am much pleased at success of column under Cookson."

The following wire has been received from Lord Kitchener, who received it from Lord Roberts :—

"DRIE KUIL, April 7th, 1902.

"Please give your brother and all under him my hearty congratulations on the success of 31st March."

By order of Lord Kitchener Bombardier Rawnsley was promoted Corporal for distinguished gallantry in the field at Boschbult.

H

1902
April 1 & 2   On April 1st The Battery marched for Korannafontein, and, after two arduous marches, reached that place on the 2nd. Here the force under General W. Kitchener halted for a few
April 6   days, and then made a forced march on Klerksdorp. The R.H.A.M.R. moved off at 7.30 p.m., reaching Leeuwkop at dawn.
April 7   The columns were then extended between Korannafontein and Leeuwkop, and a drive on Klerksdorp was made on the 7th, but only some dozen Boers were secured near Hartebeestfontein.
April 11 to 30   After a short time at Klerksdorp a column, including The Battery, was sent out to clear the country of mealies and supplies of all sorts. Rietvlei was made the centre, and the crops all round were destroyed by various methods. Many hundred wagon-loads were burnt. The work was hard and continuous.
May 1   On May 1st the force returned to Klerksdorp. It was not
May 2   till May 2nd that the Centre Section that had been serving under Colonel Rimington rejoined The Battery. It had done excellent work while detached, and it is unfortunate that no record had been kept. Lieut. Clark had been recently promoted, and Lieut. Wickham had assumed command. It was here, too, that the death of Sergt.-Farrier Dickson was first heard of. He had died of enteric at Elandsfontein on April 2nd. Throughout the campaign he had done good service; he was mentioned in despatches, and his loss was much to be deplored. The Battery took part in the " drive " from Klerksdorp to Vryburg—one of the last offensive movements of the campaign—and 365 Boers were captured. The Battery then returned to Klerksdorp, where it was at the
June 1   Declaration of Peace on June 1st.

Colonel Cookson published the following order on June 1st :—

" The following wire has been received from General Sir Ian Hamilton :—' I am ordered to revert to my post as Chief of Staff Hd. Qtrs., and must therefore bid you and your gallant troops farewell ; please convey to them my hearty good wishes and congratulations on having played so distinguished a part in the closing scene of the S. A. War, whereby I am convinced you have contributed in a very special degree to the termination of hostilities.' "

1902 Lieut. J. W. Scott and 10 N.C.O.'s and men proceeded early in the month with General W. Kitchener to collect arms from surrendered Boers.

June 20 Lieut.-Colonel Keir, R.H.A., published the following in No. 2 Column Orders :—

"KLERKSDORP, June 20th, 1902.

"Before leaving, the Officer Commanding No. 2 Column would like to express to Major Mercer and all under his command the great obligation he is under for the unfailing support he has at all times received from them. He is proud at having had under his command batteries that have not only nobly maintained the credit of their regiment on almost every field in S. Africa, but have added fame to the already brilliant record of the Royal Horse Artillery. He wishes all good-bye and God-speed.

"By order,

"(Signed)   E. S. NAIRNE, Capt., S.O.

June 22 THE BATTERY remained in camp at Klerksdorp till the 22nd, when it started for Elandsfontein.

During the month 32 N.C.O's and men left THE BATTERY time expired, and 49 joined.

July to Sept. From July to September THE BATTERY was encamped at Elandsfontein. Capt. B. Vincent joined vice Captain Sykes, invalided in July.

In September THE BATTERY was placed under orders for India. Fifty-one N.C.O.'s and men left (having less than two years to serve), and 45 joined.

Sept. 23 On September 23rd THE BATTERY entrained for Durban,
Oct. 2 where it embarked on the ss. *Ionian* on October 2nd with 33 N.C.O.'s and men who had left Bangalore in THE BATTERY.

The Officers had all changed, and altogether 17 had served, or were still serving, in THE BATTERY. The numbers by ranks were as follows :—Two majors, four captains, and 11 subalterns.

No horses were taken on board, but on leaving camp 127 were handed over. Including these, 1,088 had passed through THE BATTERY, but a large proportion had been handed over to sick-horse depôts for rest after severe marching, during the latter part of the war.

1902   THE BATTERY had fired 6,846 rounds of 15pr. and 12pr. ammunition between the February 4th, 1900, and December 31st, 1901, when it became Mounted Rifles.

The 511 rounds fired from four guns on February 9th, 1900, at Slingersfontein was the heaviest day's expenditure.

The names of those killed, wounded, or died from sickness are as follows :—

ROLL OF N.C.O'S. AND MEN KILLED IN ACTION AND DIED
OF WOUNDS.

RANK AND NAME.	DATE.	PLACE.
Sergeant S. Watson	7 Feb. 1900	Slingersfontein
Driver S. G. Hill	21 Dec. 1900	Thorndale
Gunner H. Bateman	15 Sept. 1901	Near Heilbron
Gunner J. L. Hayes	2 April 1902	Boschbolt

ROLL OF N.C.O'S. AND MEN WHO DIED OF DISEASE IN
SOUTH AFRICA.

RANK AND NAME.	DATE.	PLACE.	CAUSE.
Gunner R. Cox	26 April 1900	Bloemfontein	Enteric
Gunner B. Vallely	18 May 1900	Kronstad	Syncope
Driver W. F. Riddick	11 June 1900	Kronstad	Enteric
Driver W. Cornish	11 June 1900	Bloemfontein	Enteric
Gunner G. F. Smith	28 June 1900	Kroonstad	Enteric
Corporal T. H. Turner	1 Aug. 1900	Naaupoort	Enteric
Driver S. Snelling	30 Oct. 1900	Johannesburg	Enteric
Gunner R. Taylor	15 Nov. 1900	Pretoria	Enteric
Driver B. Yates	21 Nov. 1900	Pretoria	Enteric
Driver E. Cook	19 Feb. 1901	Rustplaats	Acute Atrophy of Liver
Driver G. Watson	26 April 1901	Newcastle	Enteric
Gunner A. Christmas	21 July 1901	Heidelberg	Morbus Cordis
Sgt.-Far. D. F. Dickson	2 April 1902	Elandsfontein	Enteric
Driver A. McInnis	15 April 1902	Klerksdorp	Enteric

Number of Men Invalided—72.

1902

ROLL OF N.C.O.'S AND MEN WOUNDED IN ACTION.

Rank and Name.	Date.	Place.	Nature, Wound.
Gunner A. E. Baker	4 June 1900	Pretoria	Shell, back and left thigh
Gunner R. McElvenna	4 June 1900	Pretoria	Shell, face and right ankle
Gunner E. J. Gover	28 Aug. 1900	Belfast	Gun shot, left foot
Gunner J. Warriner	19 Dec. 1900	Thorndale	Gun shot, knee
By.-Cl.-Mr. G. Bailey	19 Dec. 1900	Thorndale	Gun shot, elbow
Driver A. Leonard	19 Dec. 1900	Thorndale	Gun shot, neck
Driver A. G. Young	2 Mar. 1901	Paul Pieters Dorp	Assegai, thigh, accidental
Driver J. Jones	28 June 1901	Driefontein	Gun shot, knee
Driver A. Back	12 Oct. 1901	Welkom, E. T.	Gun shot, foot
By.-Whr. F. Devan	20 Jan. 1902	Near Frankfort	Gun shot, foot, accidental
Driver J. Robertson	26 Jan. 1902	Near Tafel Kop	Gun shot, leg, accidental
Driver W. E. Bartlett	29 Jan. 1902	Near Frankfort	Gun shot, neck
Gunner J. R. Knowles	23 Mar. 1902	Near Lieterberg Vlei	Gun shot, right side, accidental
Gunner H. Williams	31 Mar. 1902	Bosh Bolt.	Gun shot, leg
Driver D. Morgan	31 Mar. 1902	Bosh Bolt.	Gun shot, head

Oct. 12 The *Ionian*, with THE BATTERY on board, touched at the Seychelles for coal on October 12th, and Bombay was **Oct. 22** reached on the 22nd. The strength of THE BATTERY on landing was three officers and 157 N.C.O.'s and men.

THE BATTERY detrained at Meerut on October 31st, and was met at the station by Major-General McLeod, the whole of the Officers of the garrison, the band of the 15th Hussars, and very many others, and received a most hearty welcome. "H" Battery, Royal Horse Artillery, entertained THE BATTERY on arrival in barracks most hospitably.

Nov. THE BATTERY was unable to proceed to the Delhi manœuvres or to the Durbar, as no guns had yet been received, nor was THE BATTERY complete in horses; but 12 gunners and 14 drivers were attached to "H" Battery, which was present at both.

Dec. 28 A section, consisting of Sergts. S. Major, H. Kerry, C. Kerry, and Gunner Ryan, took part in the Durbar assault-at-arms and won the second prize in the riding and jumping competition, the only event open to British artillery and cavalry.

1903
Jan. 31 Lieut. J. W. Scott, 1 sergeant, 3 N.C.O.'s, and 8 drivers with their horses went to Agra to escort and provide teams for the Duke and Duchess of Connaught. They returned a fortnight later.

June 20 The annual dinner, which had not been held for three years owing to the War, took place at the Junior United Service Club on June 20th, Major-General A. M. Rawlins being in the chair, and 14 officers attending.

July The officers who had served in THE BATTERY during the War and Lt.-Colonel Brunker and Captain G. N. Cartwright presented the Mess with two silver statuettes representing respectively a N.C.O. in full dress and as a R.H.A. mounted rifleman.

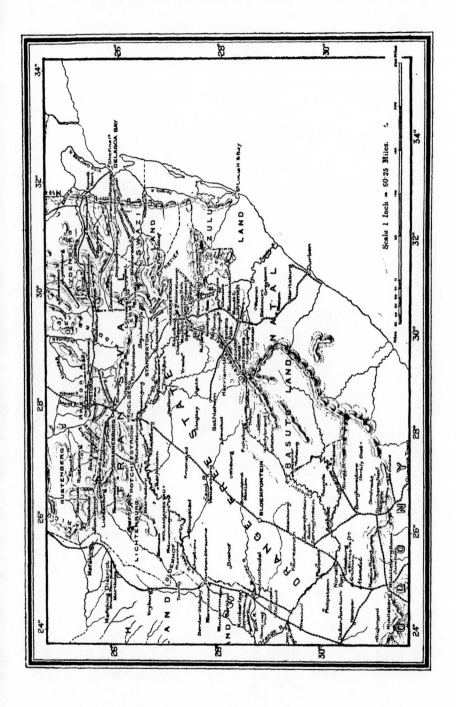

APPENDIX.

COMMANDING OFFICERS SINCE 1805.

	JOHN NOBLE, C.B.	1805
	H. T. RUDYERD	1809
	J. G. BONNER	1818
	GEORGE CONRAN	1824
5	R. S. SETON	1827
	J. WYNCH	1831
	J. T. BALDWIN	1838
	CHARLES LANCASTER	1846
	G. S. COTTER, C.B.	1852
10	JOHN D. MEIN	1854
	A. T. CADELL	1861
	G. G. B. HOLMES	1861
	N. G. CAMPBELL	1866
	E. S. MILMAN	1869
15	S. H. E. CHAMIER, C.B.	1872
	A. M. RAWLINS	1876
	W. WARD	1883
	F. S. LE GRICE	1884
	H. R. Y. BROWNE	1885
20	R. CORBETT	1885
	W. L. DAVIDSON	1891
	E. VAUGHAN-HUGHES	1896
	J. M. S. BRUNKER	1897
	P. H. ENTHOVEN	1900
25	L. H. DUCRÔT	1901

I

ENGAGEMENTS IN THE MYSORE, MAHRATTA, POLIGAR, AND JAVA WARS, AND INDIAN MUTINY, 1780—1859.

1780.	September 10.	Battle of Polipur (1).
1781.	July 1	Battle of Porto-Novo.
	August 27.	Battle of Polipur (2).
1790.	September 13.	Battle of Gazalhati Pass.
	September 16.	Battle of Velàdi.
1791.	March 7-19.	Siege of Bangalore.
	May 13.	Battle of Arikèra.
1799.	Several actions near Seringapatam.	
1801.	May 24.	Battle of Punjalum-Curchi.
	June 10.	Battle of Pàtianur.
	July 28.	Battle of Ookùr.
	July 30.	Battle of Sherewèle.
	September 18.	Battle of Peramàli.
1808.	December 28.	Battle of Amulnair.
1811.	August 10.	Battle of Welterwreeden, Java.
	August 20-26.	Siege of Cornelis, Java.
1812.	April 22.	Capture of Borang Fort
	April 25.	Occupation of Palembang.
	June 20.	Attack on the Crattan Fort and Battle of Djoecarta.
1817.	December 21.	Battle of Mahidpur.
1818.	February 27.	Siege of Talnair.
	April 17.	Battle of Sioni.
1819.	March 17 to April 9.	Siege of Assirghur.
1824.	December 4.	Battle of Kittur.
1844.	November 10.	Battle of Badraghur.
	November 30.	Battle of Purnalla.
	December 1.	Battle of Powenghur.
1858.	April 16.	Battle of Jheegum.
	April 17.	Battle of Kubrai.
	April 19.	Battle of Banda.
	April 26.	Battle of Jhounpur.
	December 25.	Relief of Kirwi.
	December 29.	Battle of Punwari and the Duddri Pass.
1859.	March 4.	Affair at Keuti.

PRINCIPAL ENGAGEMENTS IN THE SOUTH AFRICAN WAR, 1900-2.

1900.

February 4 to 11	Slingersfontein.
February 12	Worcester Kopjes.
February 20	Arundel.
February 23 and 24 ...	Kuilfontein.
March 12	Orange River.
May 10	Zand River.
May 11	Boschrand.
May 26	Vereeniging.
May 29	Johannesburg.
June 4	Pretoria.
June 11 and 12	Diamond Hill.
August 24 and 25	Belfast.
September 9	Weltevreden.
December 19	Thorndale (or Hekpoort).

1902.

February 8 ... end of the	Heilbron " Drive."
March 1 ... „	Harrismith " Drive."
March 24 ... „	Klerksdorp " Drive "
March 31	Bosh Bolt.

Names.	2nd Lieut.	Lieut.	Captain.	Major.
John Mackie	22 Aug. 1783	15 Dec. 1788	...	...
James Limond, *K.C.B.*	24 Sept. 1783	28 Sept. 1797	21 Sep. 1804	21 Oct. 1814
*C. Donaldson	8 Mar. 1784	19 May 1789	26 Feb. 1800	...
Fred Prescott	20 Oct. 1784	7 Mar. 1791	1 Mar. 1800	...
*J. Neilson	22 Oct. 1784	18 Mar. 1791	...	...
*John Noble, *C B.* ...	8 Jan. 1796	14 April 1798	21 Sep. 1804	1 Sep. 1818
Charles Gahagan ...	...	19 April 1800	11 May 1806	...
E. M. G. Showers ...	...	12 Dec. 1800	21 Sep. 1804	1 Nov. 1819
†G. W. Poignand ...	...	12 Dec. 1800	16 Feb. 1807	...
W. M. Burton... ...	...	12 Dec. 1800	24 May 1818	12 Aug. 181
F. W. Palmer ...	...	31 Mar. 1801	18 Mar. 1809	17 Jan. 1824
Walter Shairpe ...	...	22 May 1803	...	...
Hon. Thos. Rudyerd...	...	18 July 1804	9 Mar. 1810	...
Robert B. Procter ...	...	18 July 1804	6 April 1810	...
B. Macintosh ...	...	28 July 1804	24 May 1813	...
†E. J. A. Driffield ...	18 July 1804	17 Aug. 1804	...	...
J. G. Bonner	...	21 Sep. 1804	1 Jan. 1819	9 June 1825
E. S. Munro	...	14 Feb. 1805	...	...
†James Harrison ...	16 Feb. 1808	17 Feb. 1808	1 Sep. 1818	5 July 1829
†Paul Poggenpohl ...	16 Feb. 1808	17 Feb. 1808	1 Sep. 1818	...
C. R. McCausland ...	16 Feb. 1808	11 Mar. 1809	...	...
Charles W. Black ...	22 Aug. 1808	6 May 1809	14 Sep. 1819	...
Fred. Derville	11 Feb. 1809	6 April 1810	19 Oct. 1820	15 Aug. 1832
†John Jas. Gamage ...	7 July 1810	25 Nov. 1812	17 Oct. 1821	...
T. B. Y. Kennan ...	7 July 1810	24 May 1813	26 Jan. 1822	...
†George Conran ...	27 July 1811	19 May 1814	21 May 1823	1 May 1833
John Morgan Ley ...	27 July 1811	1 Sep. 1818	19 June 1824	28 Jan. 1838
John T. Kelly	2 Feb. 1812	...	...	...
Chas. H. Warre ...	11 June 1812	1 Sep. 1818	4 Sep. 1824	...
†Francis F. Whinyates...	6 July 1813	1 Sep. 1818	24 Oct. 1824	28 June 1838
Richd. Sumner Seton...	6 July 1813	1 Sep. 1818	10 June 1825	Aug. 1838

* These are the only Artillery officers whom I can trace as having served with in 1805.—*G.E.W.M.*

Lt.-Colonel.	Colonel.	General Officer.	Remarks.
... 13 May 1821	... 5 June 1829	... M.G. 28 June 1838	...
...	...	...	DIED JUNE 20, 1803
...	...	...	...
... 16 Aug. 1821	...	...	DIED —— 1793
			Commandant of the Corps of Horse Artillery, 1st February, 1809, DIED 16 JULY, 1836
...	...	...	DIED 14 APRIL, 1817
1 May 1824	2 Jan. 1833	{ M.G. 28 June 1838 / L.G. 11 Nov. 1851 / G. 4 Mar. 1858 }	} DIED 13 DECEMBER, 1868
...	...	...	DIED 10 JUNE, 1820
23 May 1825	18 June 1831	{ M.G. 23 Nov. 1841 / L.G. 11 Nov. 1851 }	Commandant of Artillery 1839. DIED 13 AUGUST, 1853
...	...	...	DIED 8 JUNE, 1825
...	...	...	DIED 25 MAY, 1809
...	...	...	DIED 24 JUNE, 1824
...	...	...	DIED 23 MAY, 1813
...	...	...	DIED 21 JANUARY, 1822
...	...	...	Mortally wounded at the battle of Welterwreeden. DIED 28 AUG., 1811
4 July 1829	10 Nov. 1843	...	Retired 4 July, 1829. DIED 3 MARCH, 1867
...	...	...	Severely wounded at the siege of Cornelis. DIED 16 JAN., 1814
...	...	...	Retired 1 May, 1833. DIED 24 JUNE, 1869
...	...	...	DIED 8 JULY, 1821
...	...	...	DIED 2 AUGUST, 1817
...	...	...	KILLED IN ACTION AT KITTOOR, 23 OCTOBER, 1824
20 Mar. 1839	3 July 1845	...	Commandant of Artillery, May, 1849. DIED 7 OCTOBER, 1849
...	...	...	Wounded at the battle of Mahidpore. DIED 27 SEPT., 1826
...	...	...	DIED 11 JUNE, 1827
26 Feb. 1840	2 Aug. 1850	{ M.G. 28 Nov. 1854 / L.G. 24 Jan. 1865 / G. 14 Dec. 1868 }	{ Commanding Horse Brigade 25 April, 1840. Commandant of Artillery, 14 Nov., 1850. DIED 28 AUGUST, 1869
23 Nov. 1841	...	...	Retired 31 December, 1842. DIED 15 FEBRUARY, 1864
...	...	...	DIED 29 JUNE, 1818
...	...	...	DIED 2 AUGUST, 1825
31 Dec. 1842	1 July 1853	{ M.G. 28 Nov. 1854 / L.G. 24 Jan. 1865 / G. 14 Dec. 1868 }	{ Commandant of Artillery, 13 March, 1851. Col.-Commandant of Horse Artillery 12 Aug., 1853. DIED 22 JANUARY, 1887
...	...	...	Retired 5 October, 1838. DIED 11 AUGUST, 1872

the Madras Galloper Guns previous to the introduction of the Regimental System
† Adjutants of THE TROOP.

Names.	2nd Lieut.	Lieut.	Captain.	Major.
John Wynch	4 July 1814	1 Sep. 1818	13 Aug. 1825	28 June 1838
William Brooke ...	4 July 1814	1 Sep. 1818	12 Nov. 1825	...
Cortlandt Taylor ...	4 July 1814	1 Sep. 1818	17 June 1826	11 Aug. 1840
Thomas Cussans ...	4 July 1814	1 Sep. 1818	...	...
†Matthew Campbell ...	16 June 1820	8 June 1821	10 June 1830	...
†C. H. Best	10 May 1822	2 Sep. 1822	28 May 1834	...
J. T. Baldwin	6 June 1823	7 June 1823	22 Dec. 1834	...
T. H. Humffreys ...	6 June 1823	25 Mar. 1824	12 April 1838	...
†John T. Ashton ...	18 Dec. 1823	1 May 1824	2 Mar. 1839	14 Oct. 1847
†A. F. Oakes	16 June 1825	28 Sep. 1826	5 Mar. 1844	11 Nov. 1851
Edward Brice, C.B. ...	16 June 1826	29 Jan. 1828	3 July 1845	20 June 1854
Charles Lancaster ...	16 June 1826	14 Mar. 1829	3 July 1845	...
J. P. Beresford ...	16 June 1826	19 April 1829	3 July 1845	28 Nov. 1854
J. E. Mawdesley ...	15 Dec. 1826	12 Oct. 1831	3 July 1845	20 June 1854
†C. J. Cooke	3 May 1827	14 Oct. 1831	13 July 1845	20 June 1854
G. S. Cotter, C.B. ...	15 June 1827	23 May 1832	24 Dec. 1845	20 June 1854
†John Moore	13 Dec. 1827	29 May 1832	11 April 1846	...
†B. W. Black	13 Dec. 1827	2 Jan. 1833	7 June 1847	20 June 1854
J. K. B. Timins ...	12 June 1828	2 Feb. 1836	...	...
W. M. Gabbett ...	12 Dec. 1828	4 June 1836	30 April 1849	1 May 1849
J. D. Mein	11 Dec. 1829	15 Oct. 1838	4 June 1852	28 Nov. 1854
Archibald Foulis ...	8 Dec. 1831	6 Jan. 1841	...	...
F. G. Nuthall	12 Dec. 1834	17 Aug. 1841	30 Aug. 1854	...
A. T. Cadell	12 Dec. 1834	17 Aug. 1841	28 Nov. 1854	...
A. V. Falls	11 June 1839	17 Aug. 1841	11 April 1855	...
G. S. Elliott	10 Dec. 1839	10 June 1842	...	...
W. C. F. Gosling ...	11 June 1840	31 Dec. 1842	9 Oct. 1856	17 Aug. 1864
G. B. B. Holmes ...	11 Dec. 1841	3 July 1845	27 April 1858	21 Dec. 1865
Robert Morton... ...	8 Dec. 1843	7 June 1847	27 Aug. 1858	...
Napier G. Campbell ...	8 Dec. 1843	2 Aug. 1848	27 Aug. 1858	...
R. G. F. Henegan ...	13 June 1845	23 Nov. 1848	27 Aug. 1858	...
G. G. J. Campbell ...	13 June 1845	13 July 1849	27 Aug. 1858	...
M. C. Lawson	13 Dec. 1845	27 June 1853	27 Aug. 1858	...
R. C. Henchy	8 June 1849	31 Dec. 1854	10 Feb. 1860	...

† Adjutants of THE TROOP

Lt.-Colonel.	Colonel·	General Officer.		Remarks.
3 July 1845	28 Nov. 1854	...		Retired 28 November, 1854. DIED 11 JANUARY, 1880
...	...	...		DIED 5 AUGUST, 1827
13 July 1845	...	...		Retired 28 August, 1846. DIED 14 JUNE, 1874
...	...	...		Wounded in the Pindari campaign. Superseded for absence without leave 23 January, 1823. DIED 16 MARCH, 1870
...	...	...		DIED 4 DECEMBER, 1831
...	...	...		Retired 20 July, 1842
...	...	...		DIED 11 APRIL, 1846
...	...	...		Retired 25 September, 1841. DIED 15 OCTOBER, 1881
...	...	...		Retired 14 October, 1847. DIED 6 DECEMBER, 1887
...	...	...		DIED AT THE CAPTURE OF RANGOON, 12 APRIL, 1852
20 July 1858	25 Sep. 1861	B.G.	25 Sep. 1861	DIED 8 JUNE, 1868
...	...	...		DIED 14 MARCH, 1854
...	...	...		Retired 28 Nov., 1854. DIED 1880
31 Dec. 1859	...	...		Retired 15 January, 1861. DIED 5 APRIL, 1874
22 Feb. 1860	1 Oct. 1862	M.G.	6 Jan. 1863	Retired 6 January, 1863. DIED 26 JANUARY, 1887
20 July 1858	1 Sep. 1860	...		Retired 25 February, 1861. DIED 25 JULY, 1878
...	...	...		Retired 26 June, 1849. DIED 20 FEBRUARY, 1883
31 Dec. 1860	6 Jan. 1863	M.G.	1 Aug. 1863	Retired 1 September, 1863. DIED 8 JANUARY, 1890
...	...	...		DIED 21 DECEMBER, 1841
28 Nov. 1854	28 Nov. 1857	...		Retired 28 August, 1861. DIED 20 JUNE, 1866
20 July 1858	8 July 1864	M.G.	1 Aug. 1872	Retired 1 August, 1872. DIED 24 NOVEMBER, 1896
...	...	...		DIED 27 JUNE, 1853
14 Sep. 1861	14 Sep. 1866	M.G.	1 Aug. 1872	Retired 1 August, 1872
25 Sep. 1861	25 Sep. 1866	M.G.	1 Oct. 1877	Retired 31 December, 1878. DIED 14 July, 1885
9 June 1862	9 June 1867	M.G.	31 Dec. 1878	Retired 31 December, 1878 DIED 21 AUGUST, 1849
17 Feb. 1865	17 Feb. 1870	L.G.	12 May 1882	DIED 14 JUNE, 1885
3 Oct. 1866	10 Nov. 1869	...		Retired 10 November, 1869. DIED 1 JUNE, 1892
31 Mar. 1869	1 Aug. 1872	...		Retired 1 August, 1872
1 Sep. 1869	1 Sep. 1874	{ L.G.	31 Mar. 1881	Col.-Commandant Royal Artillery, 6 October, 1885
		{ G.	31 Dec. 1891	
12 Sep. 1869	...	...		DIED 17 DECEMBER, 1876
...	...	...		DIED 6 JANUARY, 1865
1 Aug. 1872	...	...		Retired 1 August, 1872
1 Aug. 1872	...	...		Retired 1 August, 1872

Names.	2nd Lieut.	Lieut.	Captain.	Major.
G. C. M. Martin ...	8 June 1849	11 April 1855	30 April 1860	...
Everard S. Milman ...	11 Dec. 1849	1 May 1855	22 June 1860	...
John Stewart	14 June 1850	9 Oct. 1856	1 Sep. 1860	...
C. Johnson	13 June 1851	4 Nov. 1857	14 Sep. 1861	...
D. D. Anderson... ...	12 Dec. 1851	27 April 1858	9 June 1862	...
C. J. McMahon... ...	11 June 1853	27 April 1858	16 July 1863	5 July 1872
S. H. E. Chamier, C.B.	11 June 1853	27 April 1858	29 Feb. 1864	11 Oct. 1864
M. L. Monckton ...	13 June 1856	27 Aug. 1858	24 Mar. 1865	6 July 1872
M. H. C. B. Steinman...	12 June 1857	27 Aug. 1858	23 Nov. 1865	2 Nov. 1872
E. T. Ouchterlony ...	12 Dec. 1857	27 Aug. 1858	21 April 1866	23 Nov. 1872
H. G. Pritchard, C.S.I.	12 Dec. 1857	27 Aug. 1858	27 April 1867	12 Dec. 1877
A. M. Rawlins	12 Dec. 1857	27 Aug. 1858	6 May 1867	29 Sep. 1873
Hon. A. B. de Montmorency	12 June 1858	27 Aug. 1858	13 July 1868	3 Feb. 1876
W. R. C. Brough ...	...	11 Dec. 1858	4 July 1870	1 Oct. 1877
A. K. Seacombe ...	...	11 Dec. 1858	14 Feb. 1872	23 April 1878
W. S. Hebbert	...	10 June 1859	6 July 1872	1 July 1879
P. Sanderson, C.M.G. ...	...	9 Dec. 1859	...	...
A. I. Maclaverty ...	...	8 June 1860	1 Aug. 1872	1 Sep. 1880
M. R West	...	8 June 1860	1 Aug. 1872	1 Sep. 1880
G R. Gambier	...	8 June 1860	2 Nov. 1872	12 Dec. 1880
D. W. Lawrell	...	8 June 1860	2 Nov. 1872	26 Jan. 1881

Lt.-Colonel.	Colonel.	General Officer.	Remarks.
...	...	...	DIED 27 MAY, 1868
1 Aug. 1872	...	...	Retired 1 August, 1872
...	...	...	Retired 27 December, 1860
...	...	...	Half-Pay, 31 July, 1867. DIED 1895
...	...	...	DIED 12 JULY, 1868
27 April 1876	27 April 1881	G. 25 May 1885	Retired 29 September, 1890
17 May 1874	17 May 1879	{ M.G. 26 Feb. 1885 { L.G. 27 Oct. 1886	} Retired 27 October, 1886
...	...	...	DIED 14 JULY, 1882
...	...	...	Half-Pay, 24 May, 1877
...	...	...	DIED 25 OCTOBER, 1875
12 Dec. 1883	12 Dec. 1887	...	Transferred to Indian Staff Corps, 1871. Retired 20 November, 1893
31 July 1881	1 July 1883	M.G. 8 June 1893	Retired 19 March, 1894
4 Feb. 1883	1 Oct. 1891	...	Retired 1 October, 1896. DIED 19 MAY, 1899
31 Oct. 1884	20 May 1885	...	Retired 20 May, 1885
24 Sep. 1882	...	...	Retired 24 September, 1882
18 Nov. 1882	2 Jan. 1886	...	Retired 8 July, 1893
...	...	...	Half-Pay 8 September, 1882
27 July 1886	9 July 1889	...	Retired 8 June, 1891
19 Jan. 1888	29 Sep. 1893	...	Retired 8 June, 1894
...	...	...	DIED 9 AUGUST, 1894
1 Sep. 1883	...	...	Retired 1 September, 1883. DIED 3 AUGUST, 1884

LIST OF OFFICERS WHO HAVE SERVED IN THE BATTERY SINCE ITS TRANSFER
TO THE IMPERIAL SERVICE IN OCTOBER, 1861.

Date.	Captain.	2nd Captain.	Lieutenants.		
Oct., 1861	Cadell, A. T.	Campbell, G. G. J.	Johnson, C.	McMahon, C. J.	Pritchard, H. G.
Dec.,	Holmes, G. B. B.	Martin, G. C. M.	Steinman, M. H. C. B.	,, ,,	,, ,,
June, 1863	,, ,,	,,		,, ,,	,, ,,
July, 1864	,, ,,	Johnson, C.		deMontmorency,Hon.	,, ,,
April, 1866	,, ,,	,,	Ouchterlony, E. T.	[A. B.	
July,		Monckton, M. L.	Sanderson, P.	,, ,,	
Oct., 1867	Campbell, N. G.	,, ,,	Maclavery, A. J.	,, ,,	Gambier, G. R.
Jan.,	,, ,,	,, ,,			Brough, W. R. C.
July,	,, ,,	,, ,,		Jones, T. J.	
Dec., 1868	,, ,,	,, ,,		Seacombe, A. K.	
Oct.,	,, ,,	,, ,,	Hebbert, W. S.	Thomson, C. W.	
Nov., 1869	,, ,,	,, ,,	,, ,,		
Mar., 1870	Milman, E. S.	,, ,,	,, ,,		West, M R.
Oct.,	,, ,,	,, ,,		Gambier, G. R.	,, ,,
Sept., 1871	,, ,,	,, ,,	Hutchinson, W. L.	,, ,,	
April,					
Feb., 1872					

Date.	Major.	Captain.	Lieutenants.		
Aug., 1872	Milman, E. S.	Rawlins, A. M.	Hutchinson, W. L.	Gambier, G. R.	West, M. R.
Sept.,	Charnier, S. H. E.	,,	,,	Josselyn, J. E.	Denison, W. E.
Nov.,	,,	Brough, W. R. C.	,,	,,	,,
Nov., 1873	,,			,,	,,
April, 1874	,,	Malet, G. E. W.	Whateley, W.	,,	Beaver, P. K. L.
April, 1875	,,	,,	Dalton, J. C.	,,	,,
May, 1876	Rawlins, A. M.	,,	,,	,,	Allsopp, R.
Jan.,	,,		,,	Watson, S.	,,
Aug.,	,,	Kingscote, H. B.	Phillpotts, A. H. C.	,,	,,
Dec., 1877	,,	,,	Farmer, F. C.	Cunliffe, F. L.	,,
Aug., 1879	,,	Dolphin, H. E.	,,	,,	Maberly, C. E.
Oct.,	,,	,,	Blount, C. H.*	Fox, F. J.	,,
Mar., 1880	,,	,,	,,	,,	,,
July, 1881	Ward, W.	Tillotson, L.	,,	,,	,,
Oct., 1882	,,	,,	Thomson, J. H.	,,	Barry, J. D.
Mar., 1883	,,	Weir, H. G.	,,	,,	,,
May,	Le Grice, F. S.†	Jeffreys, H. B.	,,	Baillie, A. C.	,,
Oct.,	,,	,,	,,	,,	Dames, F. T. L.
June, 1884	,,	,,		,,	,,
Sept.,	Browne, H. R. Y.				,,
Nov.,	Corbett, R.				
Jan., 1885					
June,					
Dec.,					

* Died, De Aar, S. Africa, 1900.   † Died 1902.

LIST OF OFFICERS WHO HAVE SERVED IN THE BATTERY—*Continued.*

Date.	Major.	Captain.	Lieutenants.		
April, 1886	Corbett, R.	Jeffreys, H. B.	Thomson, J. H.	Askwith, J. B. H.	Dames, F. T. L.
April, 1887	"	"	Fairholme, W. E.	"	Chamier, G. D.
May,	"	"	Calley, H.*	"	"
Oct., 1888	"	Lindsay, W. F. L.	"	"	"
Feb.,	"	"	"	"	Smith, Edmund P.
April, 1889	"	"	Wray, J. C.	"	Paget, V. F. W. A.
July,	"	"	"	"	Heath, F. W.
April, 1890	"	"	"	Elton, F. A. G. Y.	"
May,	"	"	"	"	"
Sept., 1891	Davidson, W. L	Taylor, P. B.	"	"	"
Feb., 1892	"	"	"	"	Williams, A. S.
June, 1893	Vaughan-Hughes, E.†	Connal, A. C.‡	Budworth, C. E. D.	Hobson, J. A.	"
April, 1895	"	"	"	"	"
Mar., 1896	Brunker, J. M. S.	"	"	"	Cartwright, G. N.
Oct.,	"	"	"	"	"
Mar., 1897	"	Corbyn, H.	"	"	Hon. G. F. Stanley
July,	"	"	"	Hine-Haycock, V. R.	"
Dec., 1898	"	"	Broughton, L. H. D.	"	"
Jan., 1899	"	"	"	"	"
Feb.,	Enthoven, P. H.	"	"	"	Monkhouse, W. P.
May,	"	"	"	"	"
Oct.,	"	"	"	"	"
Jan., 1900	"	"	"	"	Wood, C. H.
May,	"	"	"	"	"

* Died November 24, 1898.   † Died August 30, 1899.   ‡ Died February 2, 1899.

LIST OF OFFICERS WHO HAVE SERVED IN THE BATTERY—*Continued.*

DATE.	MAJOR.	CAPTAIN.	LIEUTENANTS.		
July, 1900	Enthoven, P. H.	Arbuthnot, D.	Broughton, L. H. D.	Hine-Haycock, V. R.	Wood, C. H.
Oct.,	"	Sykes, C. A.	Hill, H. W.	R. T. Hill	Clark, W. E.
Dec.,	Ducrot, L. H.	"	"	"	"
Mar., 1901	"	"	Scott, J. W.	Hartland-Mahon, M.	"
June,	"	"	"	Harris, O. M. [J. C.]	Wickham, T. E. P.
April, 1902	"	Vincent, B.		"	"
May,					
June,					

Printed in the United Kingdom
by Lightning Source UK Ltd.
108984UKS00002B/76